Mathematically Speaking
A Dictionary of Quotations

About the Compilers

Carl C Gaither was born on 3 June 1944 in San Antonio, Texas. He has conducted research work for the Texas Department of Corrections and for the Louisiana Department of Corrections. Additionally he has worked as an Operations Research Analyst for the past ten years. He received his undergraduate degree (Psychology) from the University of Hawaii and has graduate degrees from McNeese State University (Psychology), North East Louisiana University (Criminal Justice), and the University of Southwestern Louisiana (Mathematical Statistics).

Alma E Cavazos-Gaither was born on 6 January 1955 in San Juan, Texas. She has previously worked in quality control, material control, and as a bilingual data collector. She received her associate degree (Telecommunications) from Central Texas College and is presently attending Central Texas College and working toward a major in Business Management with a minor in Spanish.

Together they selected and arranged quotations for the books *Statistically Speaking: A Dictionary of Quotations* (Institute of Physics Publishing, 1996) and *Physically Speaking: A Dictionary of Quotations on Physics and Astronomy* (Institute of Physics Publishing, 1997).

About the Illustrator

Andrew Slocombe was born in Bristol in 1955. He spent four years of his life at Art College where he attained his Honours Degree (Graphic Design). Since then he has tried to see the funny side to everything and considers that seeing the funny side to mathematics has tested him to the full! He would like to thank Carl and Alma for the challenge!

The photograph of the authors on the previous page was taken by Kerry Yancey.

Mathematically Speaking
A Dictionary of Quotations

Selected and Arranged by

Carl C Gaither
and
Alma E Cavazos-Gaither

Illustrated by Andrew Slocombe

Institute of Physics Publishing
Bristol and Philadelphia

IOP Publishing Ltd has attempted to trace the copyright holders of all the quotations reproduced in this publication and apologizes to copyright holders if permission to publish in this form has not been obtained.

British Library Cataloguing-in-Publication Data
A catalogue record for this book is available from the British Library.

ISBN 0 7503 0503 7

Library of Congress Cataloging-in-Publication Data are available

Published by Institute of Physics Publishing, wholly owned by The Institute of Physics, London

Institute of Physics Publishing, Dirac House, Temple Back, Bristol BS1 6BE, UK
US Office: Institute of Physics Publishing, Suite 1035, The Public Ledger Building, 150 South Independence Mall West, Philadelphia, PA 19106, USA

Typeset in TEX using the IOP Bookmaker Macros
Printed in Great Britain by J W Arrowsmith Ltd, Bristol

To the bright future of our children
Mrs Melisa M Donavan, Marcus A Sutton, Matthew C Cavazos,
Maritza M Cavazos, Maurice Moore II, Marilynn D Moore,
Carl C Gaither III, H Wayne Gaither, and Russell J Gaither

and

IN LOVING MEMORY
of

Rosa "Rosie" Maria Cervantes
(7 March 1952 – 17 October 1997)
Sister & Friend

Por Fin
by
"Rosie"

Por fin – ahora soy feliz
Por fin – e realizado el amor
Soñado en mi corazon
Seras como una bendicion
Calmaste tu mis penas – que
era una condena y una maldicion
Ahora – se acaba mi sufrir
Mi alma a vuelto a ser feliz
Por fin – Ahora soy feliz
Por fin e realizado el amor soñado en
mi corazon. Seras como una bendicion
Calmaste tu mis penas que era una
condena y una maldicion
Ahora – se acaba mi sufrir
Mi alma a vuelto a ser feliz – Por fin – ahora soy feliz
Por fin e realizado el amor soñado en
mi corazon.

But chief, thou Nurse of the Didactic Muse,
Divine NONSENSIA, all thy soul infuse;
How loves and Graces in an *Angle* dwell;
How slow progressive *Points* protract the *Line*,
As pendent spiders spin the filmy twine;
How lengthened *Lines*, impetuous sweeping round,
Spread the wide *Plane*, and mark its circling bound;
How *Planes*, their substance with their motion grown,
Form the huge *Cube*, the *Cylinder*, the *Cone*.

Frere, C.
Canning, B.
In Charles Edmund's
Poetry of the Anti-Jacobian
The Loves of the Triangle
Canto I, l. 35–44

CONTENTS

PREFACE

Lichtenberg (quoted in Paul Davies' *About Time* (p. xvii)) said

A writer who wishes to be read by posterity must not be averse to putting hints which might give rise to whole books, or ideas for learned discussions, in some corner of a chapter so that one should think he can afford to throw them away by the thousand.

We have extracted and written on the following pages a great number of hints which we have found in the books we have reviewed. This book has compiled the hints and ideas from a great quantity of books for you, the reader, to enjoy and use in your daily activities.

Dictionaries are like watches; the worst is better than none, and the best cannot be expected to go quite true. Thus Doctor Samuel Johnson wrote in the preface to his Dictionary. All of the contents of this book have been compiled from widely scattered sources. They are presented in this book for the first time in a form that is readily accessible for pleasure or for reference. Although there are doubtless errors in the quotations and in the citations, and there is certainly much omission, we have sought to ease the reader's need to undertake long searches of the literature in mathematics, literature, and philosophy.

Unto those Three Things which the Ancients held impossible, there should be added this Fourth, to find a Book Printed without erratas. Alfonso de Cartagena is reported to have made this statement and toward this end we respond with William Kitchner's insight:

> That this book has its fault, no one can doubt,
> Although the Author could not find them out.
> The faults you find, good Reader, please to mend,
> Your comments to the Author kindly send.
>
> *The Economy of the Eyes*
> Part II

Mathematically Speaking is a book of quotations. It has, for the first time, brought together in one easily accessible form the best expressed

xi

thoughts that are especially illuminating and pertinent to the discipline of mathematics. Some of the quotations are profound, others are wise, some are witty, but none are frivolous. Quotations from the most famous men and women lie in good company with those from unknown wits. You may not find all the quoted 'jewels' that exist, but we are certain that you will find a great number of them here. We believe that Benjamin Franklin was correct when he said that "Nothing gives an author so much pleasure as to find his work respectfully quoted...".

Mathematically Speaking is also an aid for the individual who loves to quote—and to quote correctly. "Always verify your quotations" was advice given to Dean John William Bourgen, then fellow of Oriel College, by Dr Martin Joseph Routh. That advice was given over 150 years ago and today it is still true. Frequently, books on quotations will have subtle changes to the quotation, changes to punctuation, slight changes to the wording, even misleading information in the attribution, so that the compiler will know if someone used a quotation from 'their' book. We attempted to verify each and every one of the quotations in this book to ensure that they are correct.

The attributions give the fullest possible information that we could find to help you pinpoint the quotation in its appropriate context or discover more quotations in the original source. Speeches include, when possible, the date of the speech. We assure the reader that not one of the quotations in this book was created by us.

In summary, *Mathematically Speaking* is a book that has many uses. You can:

- Identify the author of a quotation.
- Identify the source of the quotation.
- Check the precise wording of a quotation.
- Discover what an individual has said on a subject.
- Find sayings by other individuals on the same subject.

How to Use This Book

1. A quotation for a given subject may be found by looking for that subject in the alphabetical arrangement of the book itself. To illustrate, if a quotation on "tensors" is wanted, you will find nine quotations listed under the heading tensors. The arrangement of quotations in this book under each subject heading constitutes a collective composition that incorporates the sayings of a range of people.
2. To find all the quotations pertaining to a subject and the individuals quoted use the SUBJECT BY AUTHOR INDEX. This index will help guide you to the specific statement that is sought. A brief extract of each quotation is included in this index.
3. If you recall the name appearing in the attribution or if you wish to read all of an individual author's contributions that are included in this book then you will want to use the AUTHOR BY SUBJECT

INDEX. Here the authors are listed alphabetically along with their quotations. The birth and death dates are provided for the authors whenever we could determine them.

Thanks

It is never superfluous to say thanks where thanks are due. I would first like to extend my thanks to my stepdaughter Maritza Marie Cavazos for her assistance in tracking down incomplete citations, looking for books in the libraries, and helping to sort the piles of correspondence generated in obtaining permissions. I thought that she would have been burnt out after working so hard on the previous two books *Statistically Speaking* and *Physically Speaking*. But she didn't. Next we send our deep appreciation to Jim Revill our publisher and Al Troyano our production editor, of Institute of Physics Publishing, who have assisted us so very much.

The following libraries are acknowledged for allowing us to use their resources: The Jesse H. Jones Library and the Moody Memorial Library, Baylor University; the A. Frank Smith, Jr. Library Center, Southwestern University; the main library of the University of Mary-Hardin Baylor; the main library of the Central Texas College; the Perry-Castañeda Library, the Undergraduate Library, the Engineering Library, the Law Library, Physics-Math-Astronomy Library, and the Humanities Research Center of the University of Texas at Austin.

We wish also to thank Brian Camp, Joe Gonzalez, Laura Izzo, Robert Clontz, Kenneth McFarland, Chris Braun, Lisa Sparkman, Nancy Wiles, and Paul Linkletter, and all the other circulation staff at the Perry-Castañeda Library, and Molly White, Librarian, Jason Overby, Kate Cunningham, Peter Bernacki, and Crawford Spooner, and all the other circulation staff at the Physics-Math-Astronomy Library for putting up with us when we were checking out and returning the large numbers of library books. Only they know the actual numbers.

Additionally, we would like to thank each of the publishers who provided permission to use the quotations. We made a very serious attempt to contact the publishers for permission to use the quotations. Letters were written to each publisher or agent for which we could find an address. A follow-up letter was sent to those who did not respond to our first letter. If no response was received we then assumed a calculated risk and incorporated the quotation. In no way did we use a quotation without attempting to obtain prior approval.

Carl Gaither
Alma Cavazos-Gaither
February 1998

ABSTRACTION

Whitehead, Alfred North
. . . to be an abstraction does not mean that an entity is nothing. It merely means that its existence is only one factor of a more concrete element of nature.

The Concept of Nature
Summary (p. 171)

The point of mathematics is that in it we have always got rid of the particular instance, and even of any particular sorts of entities. So that for example, no mathematical truths apply merely to fish, or merely to stones, or merely to colours. So long as you are dealing with pure mathematics, you are in the realm of complete and absolute abstraction . . .. Mathematics is thought moving in the sphere of complete abstraction from any particular instance of what it is talking about.

Science and the Modern World
Mathematics as an Element in the History of Thought (pp. 31–2)

ADDITION

Carroll, Lewis
"What's one and one and one and one and one and one and one and one and one and one?"

"I don't know," said Alice. "I lost count."

"She can't do Addition," the Red Queen interrupted.

The Complete Works of Lewis Carroll
Through the Looking Glass
Queen Alice

Dickson, Paul
Baldy's Law. Some of it plus the rest of it is all of it.

The Official Rules (p. B-8)

Dostoevsky, Fyodor Mikhailovich
Twice two makes four seems to me simply a piece of insolence. Twice two makes four is a pert coxcomb who stands with arms akimbo barring your path and spitting. I admit that twice two makes four is an excellent thing, but if we are to give everything its due, twice two makes five is sometimes a very charming thing too.

The Short Novels of Dostoevsky
Notes From Underground (p. 139)

Hardy, Thomas
. . . is a woman a thinking unit at all, or a fraction always its integer? How do you argue that marriage was only a clumsy contract—which it is—how you showed all the objections to it—all the absurdities! If two and two made four when we were happy together, surely they make four now? I can't understand it, I repeat!

Jude the Obscure
Part Sixth, Chapter 3
At Christminster Again (p. 359)

2

Housman, Alfred Edward
To think that two and two are four
  And neither five nor three,
The heart of man has long been sore
  And long 'tis like to be.

The Collected Poems of A.E. Housman
Last Poems
XXXV (p. 142)

Ice-T
I write rhymes with addition and algebra, mental geometry.

O.G.: Original Gangster CD
Mind Over Matter

Pope, Alexander
Ah! why, ye Gods! Should two and two make four?

The Complete Poetical Works
The Dunciad
Book 2, l. 285

Shaw, George Bernard
MRS. BASHOM: At school I got as far as addition and subtraction; but I never could do multiplication or division.

NEWTON: Why, neither could I: I was too lazy. But they are quite unnecessary: addition and subtraction are quite sufficient. You ask the logarithms of the numbers; and the antilogarithm of the sum of the two is the answer. Let me see: three times seven?

The Complete Plays of Bernard Shaw
In Good King Charles's Golden Days
Act I (p. 1335)

Unknown
There was an old man who said, "Do
Tell me *how* I should add two and two?
  I think more and more
  That it makes about four—
But I fear that is almost too few."

Reproduced in Helen Plotz
Imagination's Other Place
There was an Old Man Who Said, "Do (p. 79)

One and one make two,
But if one and one should marry,
  Isn't it queer
  Within a year
There's two and one to carry.

Source unknown

Si inaequalibus aequalia addas, omnia errunt inaequalia.
[If equals be added to unequals, the whole will be unequal.]

Source unknown

$1 + 1 = 3$, for large values of 1.

Source unknown

West, Mae
One figure can sometimes add up to a lot.

The Wit and Wisdom of Mae West (p. 35)

I learned that two and two are four and five will get you ten if you know how to work it.

The Wit and Wisdom of Mae West (p. 52)

ALGEBRA

Barrie, James Matthew
PHOEBE: Algebra! It—it is not a very ladylike study Isabella.

The Plays of J.M. Barrie
Quality Street
Act II (p. 113)

MISS SUSAN: What is algebra exactly; is it those three-cornered things?

PHOEBE: It is *x* minus *y* equals *z* plus *y* and things like that. And all the time you are saying they are equal, you feel in your heart, why should they be?

The Plays of J.M. Barrie
Quality Street
Act II (p. 115)

Brahmagupta
As the sun eclipses the stars by his brilliancy, so the man of knowledge will eclipse the fame of others in assemblies of the people if he proposes algebraic problems, and still more if he solves them.

Quoted in Florian Cajori
A History of Mathematics
The Hindus (p. 92)

Butler, Samuel
And wisely tell what hour o' th' day
The clock does strike by Algebra.

The Poetical Works of Samuel Butler
Volume I
Hudibras
Part I, Canto I, l. 125

Cajori, Florian
The best review of arithmetic consists in the study of algebra.

The Teaching and History of Mathematics in the United States (p. 110)

5

Carroll, Lewis
Let U = the University, G = Greek, and P = Professor, then GP = Greek
Professor . . .

The Complete Poetical Works of Lewis Carroll
The New Method of Evaluation (p. 1124)

Clifford, William Kinston
We may always depend upon it that algebra, which cannot be translated
into good English and sound common sense, is bad algebra.

Common Sense in the Exact Sciences
Chapter 1, section 7 (p. 20)

Cochran, William G.
Cox, Gertrude M.
. . . polynomials are notoriously untrustworthy when extrapolated.

Experimental Designs (p. 336)

Comte, Auguste
Algebra is the *Calculus of Functions*, and Arithmetic the *Calculus of Values*.

Philosophy of Mathematics (p. 56)

Dantzig, Tobias
The word is of Arabic origin. *"Al"* is the Arabic article *the*, and *"gebar"*
is the verb *to set*, to restitute.

Number: The Language of Science (p. 76)

Date, J.C.B.
To Algebra God is inclined—
The world is a thought in His Mind.
  It seems so erratic,
  Because it's quadratic,
And the roots are not easy to find.

Quoted in E.O. Parrott (Editor)
The Penguin Book of Limericks
Theory and Practice

de Fontenelle, Bernard
Nothing proves more clearly that the mind seeks truth, and nothing
reflects more glory upon it, than the delight it takes, sometimes in spite
of itself, in the driest and thorniest researches of algebra.

Histoire du Renouvellement de l'Académie des Sciences
Preface

de Morgan, Augustus
The first thing to be attended to in reading any algebraical treatise, is the gaining a perfect understanding of the different processes there exhibited, and of their connection with one another. This cannot be attained by a mere reading of the book, however great the attention which may be given. It is impossible, in a mathematical work, to fill up every process in the manner in which it must be filled up in the mind of the student before he can be said to have completely mastered it. Many results must be given of which the details are suppressed, such are the additions, multiplications, extractions of the square root, etc., with which the investigations abound. These must not be taken on trust by the student, but must be worked by his own pen, which must never be out of his hand, while engaged in any algebraical process.

On the Study and Difficulties of Mathematics (pp. 175–6)

Einstein, Jacob
Algebra is a merry science.

Quoted in Ronald W. Clark
Einstein: The Life and Times (p. 12)

Johnson, Samuel
. . . I would advise you Sir, to study algebra, if you are not an adept already in it . . .

Johnsonian Miscellanies
Volume 1
Anecdotes (p. 301)

Langer, Susanne K.
Behind these symbols lie the boldest, purest, coolest abstractions mankind has ever made. No schoolman speculating on essences and attributes ever approached anything like the abstractness of algebra.

Philosophy in a New Key
Chapter 1 (p. 18)

Lebowitz, Fran
Stand firm in your refusal to remain conscious during algebra. In real life, I assure you, there is no such thing as algebra.

Social Studies
Tips for Teens (p. 36)

Lipsky, Eleazar
. . . Liberty and Justice for All. This principle can be represented as follows:

$$L + J = 1$$

where L = Liberty; J = Justice; and 1 = All.

It is evident that as J increases in value, L will decrease. In simple terms, the more Justice, the less Liberty. This is almost a truism . . . The formula was modified to

$$\frac{J}{L} = 1$$

or:

$$J = 1(L)$$

from which it follows, the more Justice, the *more* Liberty.

The Magazine of Fantasy and Science Fiction
Snitkin's Law
February 1959 (p. 61)

Locke, John
They that are ignorant of Algebra cannot imagine the wonders in this kind are to be done by it: and what further improvements and helps advantageous to other parts of knowledge the sagacious mind of man may yet find out, it is not easy to determine. This at least I believe, that the *ideas of quantity* are not those alone that are capable of demonstration and knowledge; and that other, and perhaps more useful, parts of contemplation, would afford us certainty, if vices, passions, and domineering interest did not oppose and menace such endeavors.

An Essay Concerning Human Understanding
Book 4, Chapter 3, section 18

Mach, Ernst
The object of all arithmetical operations is to *save* direct enumeration, by utilizing the results of our old operations of counting. Our endeavor is, having done a sum once, to preserve the answer for future use . . .. Such, too, is the purpose of algebra, which, substituting relations for values, symbolizes and definitely fixes all numerical operations which follow the same rule.

The Science of Mechanics
Chapter IV, section iv, part 5 (p. 585)

Morley, Christopher
Marriage is the square of a plus b
In other words
$a^2 + b^2 + 2ab$
Where $2ab$ (of course)
Are twins.

Translations from the Chinese
$(a+b)^2$

Moser, Leo
A quadratic function, ambitious,
Said, 'It's not only wrong, but it's vicious.
   It's surely no sin
    To have max. and min.;
To limit me so is malicious'.

Quoted in E.O. Parrott (Editor)
The Penguin Book of Limericks
Theory and Practice

Pastan, Linda
I used to solve equations easily.
If train A left Sioux Falls
at nine o'clock, traveling
at a fixed rate,
I knew when it would meet train B.
Now I wonder if the trains will crash;
or else I picture naked limbs
through Pullman windows, each
a small vignette of longing.

Quoted in Ernest Robson and Jet Wimp
Against Infinity
Algebra (p. 50)

Trumbull, John
What though in algebra, his station
Was negative in each equation . . .

Quoted in Florian Cajori
The Teaching and History of Mathematics in the United States (p. 62)

Unknown
Algebra begins with the unknown and ends with the unknowable.

Source unknown

A young man, to impress his girl friend: "I'm taking four courses at the university—German, French, Russian and Algebra."

His girl friend replied: "Gee, you're a genius! Now, darling, do tell me 'I love you' in Algebra."

Source unknown

When you long for the good old days of your youth, just think of algebra.

Source unknown

The human mind has never invented a labor-saving machine equal to algebra.

The Nation
Two Works of Algebra
Volume 33, Number 847, September 22, 1881 (p. 237)

Does it lie 'mid Algebra's stern array,
Where the Law of Symmetry points the way,
And the path leads up through ascending powers
To the hilltop won after weary hours?

The Mathematical Gazette
The Happy Land
Volume VIII, Number 117, May 1915 (p. 99)

After applying some simple algebra to some trite phrases and clichés a new understanding can be reached of the secret to wealth and success.

Knowledge is Power, Time is Money and as every engineer knows, Power is Work over Time.

So, substituting algebraic equations for these time worn bits of wisdom, we get:

$$K = P \qquad (1)$$

$$T = M \qquad (2)$$

$$P = W/T \qquad (3)$$

Now, do a few simple substitutions:

Put W/T in for P in equation (1), which yields:

$$K = W/T. \qquad (4)$$

Put M in for T into equation (4), which yields:

$$K = W/M. \qquad (5)$$

Now we've got something. Expanding back into English, we get:

Knowledge equals Work over Money.

What this MEANS is that:

1. The More You Know, the More Work You Do, and
2. The More You Know, the Less Money You Make.

Solving for Money, we get:

$$M = W/K \qquad (6)$$

Money equals Work Over Knowledge.

From equation (6) we see that Money approaches infinity as Knowledge approaches 0, regardless of the Work done.

What THIS MEANS is:

The More you Make, the Less you Know.

Solving for Work, we get

$$W = MK \tag{7}$$

Work equals Money times Knowledge.

From equation (7) we see that Work approaches 0 as Knowledge approaches 0. What THIS MEANS is:

The stupid rich do little or no work.

Mathematical Wisdom
Source unknown

Whitehead, Alfred North
. . . algebra is the intellectual instrument which has been created for rendering clear the quantitative aspects of the world.

The Organization of Thought (pp. 14–15)

ANALOGY

Banach, Stefan
Good mathematicians see analogies between theorems or theories, the very best ones see analogies between analogies.

> Quoted in S.M. Ulam
> *Adventures of a Mathematician*
> Chapter 10 (p. 203)

Cross, Hardy
An analogy is not a reason . . .

> *Engineers and Ivory Towers*
> For Man's Use of God's Gifts (p. 109)

Hodnett, Edward
Analogy suggests rather than proves.

> *The Art of Problem Solving*
> Analogy (p. 143)

Latham, Peter Mere
It is safest and best to fill up the gaps of our knowledge from analogy.

> Quoted in William B. Bean
> *Aphorisms from Latham* (p. 37)

Melville, Herman
O Nature, and O soul of man! how far beyond all utterance are your linked analogies! not the smallest atom stirs or lives on matter, but has its cunning duplicate in mind.

> *Moby Dick*
> The Sphinx

ARITHMETIC

Brandeis, Louis D.
. . . obsessed with the delusion that two and two make five, he fell at last a victim of the relentless rules of humble arithmetic. Remember, O Stranger! Arithmetic is the first of the sciences and the mother of safety.

Quoted in Alpheus Mason
Brandeis: A Free Man's Life (p. 200)

Brownell, William A.
The arithmetic of 1900 differed materially from the arithmetic we now include in the elementary curriculum . . . . Two of its characteristics stand out prominently: (a) it was hard, and (b) it was little related to practical living . . . .. Teachers, relying pretty much upon what was in the textbook, showed pupils what to do and then relied upon abundant bodies of practice to produce mastery. Homework assignments were heavy, and many parents were called upon to revive, temporarily at least, skills that they had forgotten. The children who survived this demanding regimen, aided often . . . by feats far beyond the capabilities of eighth-graders today, whether or not they ever later put them to effective use.

The Arithmetic Teacher
The Revolution in Arithmetic
Volume 1, Number 1, February 1954 (p. 1)

Byron, Lord George Gordon
I know that two and two make four—& should be glad to prove it too if I could—though I must say if by any sort of process I could convert 2 & 2 into *five* it would give me much greater pleasure.

Quoted in Leslie A. Marchand (Editor)
Byron's Letters and Journals
Volume 3, Number 10
Letter to Annabella Milbanke
November 10th, 1813 (p. 159)

Carroll, Lewis
. . . the different branches of Arithmetic—Ambition, Distraction, Uglification, and Derision.

The Complete Works of Lewis Carroll
Alice's Adventures in Wonderland
The Mock Turtle's Story

Churchill, Winston Spencer
You cannot ask us to take sides against arithmetic.

Quoted in F.B. Czarnomski
The Wisdom of Winston Churchill (p. 44)

Dewey, John
The way to enable a student to apprehend the instrumental value of arithmetic is not to lecture him on the benefit it will be to him in some remote and uncertain future, but to let him discover that success in something he is interested in doing depends on ability to use numbers.

Democracy and Education
Chapter XVIII (p. 281)

Emerson, Ralph Waldo
It is better to teach the child arithmetic and Latin grammar than rhetoric and moral philosophy, because they require exactitude of performance; it is made certain that the lesson is mastered, and that power of performance is worth more than the knowledge.

The Portable Emerson
Education (pp. 262–3)

For we do not listen with the best regard to the verses of a man who is only a poet, nor to his problems if he is only an algebraist; but if a man is at once acquainted with the geometric foundations of things and with their festal splendor, his poetry is exact and his arithmetic musical.

Society and Solitude
Work Days (p. 511)

Heinlein, Robert A.
An intellectual is a highly educated man who can't do arithmetic with his shoes on, and is proud of his lack.

The Cat Who Walks Through Walls (p. 393)

Holmes, Oliver Wendell
I was just going to say, when I was interrupted, that one of the many ways of classifying minds is under the heads of arithmetical and algebraical intellects. All economical and practical wisdom is an extension or variation of the following arithmetical formula: 2 + 2 = 4.

Every philosophical proposition has the more general character of the expression $a + b = c$. We are the mere operatives, empiric, and egoists, until we learn to think in letters instead of figures.

The Autocrat of the Breakfast-Table
Chapter I

Hoyle, Fred
Once I had learnt my twelve time table (at the age of three); it was down hill all the way.

Source unknown

Kasner, Edward
Newman, James R.
Arithmetic has been the queen and the handmaiden of the sciences from the days of the astrologers of Chaldea and the high priests of Egypt to the present days of relativity, quanta, and the adding machine. Historians may dispute the meaning of ancient papyri, theologians may wrangle over the exegesis of Scripture, philosophers may debate over the Pythagorean doctrine, but all will concede that the numbers in the papyri, in the Scriptures and in the writings of Pythagoras are the same as the numbers of today. As arithmetic, mathematics has helped the risings of the Nile, to measure fields and the height of the Pyramids, to measure the speed of a stone as it fell from a tree in Woolsthorpe, to weigh the stars and the atoms, to mark the passage of time, to find the curvature of space. And although mathematics is also the calculus, the theory of probability, the matrix algebra, the science of the infinite, it is still the art of counting.

Mathematics and the Imagination (p. 28)

Kipling, Rudyard
A rule to trick th' arithmetic.

Rudyard Kipling's Verse
To the True Romance

La Touche, Mrs.
I do hate sums. There is no greater mistake than to call arithmetic an *exact* science. There are Permutations and Aberrations discernible to minds entirely noble like mine; subtle variations which ordinary accountants fail to discover; hidden laws of Numbers which it requires a mind like mine to perceive. For instance, if you add a sum from the bottom up, and then again from the top down, the result is always different.

Mathematical Gazette
Gleaning Far and Near
Volume 12, Number 170, May 1924 (p. 95)

Leacock, Stephen
His brain trained by long years of high living and plain thinking had become too subtle, too refined an instrument for arithmetic . . .

Literary Lapses
Lord Oxhead's Secret (p. 17)

The student of arithmetic who has mastered the first four rules of his art, and successfully striven with money sums and fractions, finds himself confronted by an unbroken expanse of questions known as problems.

Literary Lapses
A, B, and C (p. 118)

Leibniz, Gottfried Wilhelm
The pleasure we obtain from music comes from *counting*, but counting unconsciously. Music is nothing but unconscious arithmetic.

Quoted in Oliver Sacks
The Man Who Mistook His Wife for a Hat and Other Clinical Tales
The Twins (p. 195)

Lieber, Lillian R.
In other words, without a theory, a plan, the mere mechanical manipulation of the numbers in a problem does not necessarily make sense just because you are using Arithmetic!

The Education of T.C. MITS (p. 36)

Mikes, George
A English professor of mathematics would say to his maid adding up his shopping list: "I'm no good at arithmetic, I'm afraid. Please correct me, Jane, if I am wrong, but I believe that the square root of 97,344 is 312."

How to Be an Alien
Britain (p. 14)

Nash, Ogden
. . . the only way I can distinguish proper from improper fractions
Is by their actions.

Parents Keep Out
Ask Daddy, He Won't Know

Nicomachus of Gerasa
. . . arithmetic . . . which is the mother of geometry . . .

Introduction to Arithmetic
Translation Book II
Chapter VI, 1 (p. 237)

Parker, F.W.
The science of Arithmetic may be called the science of exact limitation of matter and things in space, force, and time.

Talks on Pedagogics (p. 64)

Plato
SOC.: . . . if arithmetic, mensuration, and weighing be taken away from any art, that which remains will not be much.

Philebus
Section 55

SOC.: "And so, Gorgias, you call arithmetic rhetoric." But I do not think that you really call arithmetic rhetoric any more than geometry would be so called by you.

Gorgias
Section 450

. . . arithmetic has a very great and elevating effect, compelling the soul to reason about abstract number, and rebelling against the introduction of visible or tangible objects into the argument.

The Republic
Book VII, Section 525

. . . those who have a natural talent for calculation are generally quick at every other kind of knowledge; and even the dull, if they have had an arithmetical training, although they may derive no other advantage from it, always become much quicker than they would otherwise have been . . .

The Republic
Book VII, Section 526

. . . arithmetic is a kind of knowledge in which the best natures should be trained, and which must not be given up.

The Republic
Book VII, Section 526

Rosenblatt, Roger
. . . Uncle Scrooge preferred to let the poor die "and decrease the surplus population." Scrooge may not have God on his side, but his arithmetic was impeccable.

The Man in the Water
Do You Feel the Deaths of Strangers? (p. 177)

Sandburg, Carl
Arithmetic is where numbers fly like pigeons in and out of your head.

Harvest of Poems 1910–1960
Arithmetic

Arithmetic is numbers you squeeze from your head to your hand to your pencil to your paper till you get the answer.

Harvest of Poems 1910–1960
Arithmetic

Smith, H.J.S.
[Arithmetic] is one of the oldest branches, perhaps the very oldest branch, of human knowledge; and yet some of its most abstruse secrets lie close to its tritest truths.

Quoted in E.T. Bell
Men of Mathematics (p. xv)

Smith, Sydney
What would life be like without arithmetic, but a scene of horrors.

The Letters of Sydney Smith
Letter 692
To Miss Lucie Austin, 22 July 1835 (p. 622)

Steinbeck, John
He was an arithmetician rather than a mathematician. None of the humour, the music, or the mysticism of higher mathematics ever entered his head. Men might vary in height or weight or colour, just as 6 is different from 8, but there was little other difference.

The Moon is Down
Chapter Two (p. 22)

Unknown
Arithmetically speaking, rabbits multiply faster than adders add.

Source unknown

White, E.B.
The fifth-graders were having a lesson in arithmetic, and their teacher, Miss Annie Snug, greeted Sam with a question.

"Sam, if a man can walk three miles in one hour, how many miles can he walk in four hours?"

"It would depend on how tired he got after the first hour," replied Sam.

The other pupils roared. Miss Snug rapped for order.

"Sam is quite right," she said. "I never looked at the problem that way before. I always supposed that man could walk twelve miles in four hours, but Sam may be right: that man may not feel so spunky after the first hour. He may drag his feet. He may slow up."

Albert Bigelow raised his hand. "My father knew a man who tried to walk twelve miles, and he died of heart failure," said Albert.

"Goodness!" said the teacher. "I suppose *that* could happen, too."

"Anything can happen in four hours," said Sam. "A man might develop a blister on his heel. Or he might find some berries growing along the road and stop to pick them. That would slow him up even if he wasn't tired or didn't have a blister."

"It would indeed," agreed the teacher. "Well, children, I think we have all learned a great deal about arithmetic this morning, thanks to Sam Beaver."

Everyone had learned how careful you have to be when dealing with figures.

The Trumpet of the Swan
School Days (pp. 63–4)

I do hate sums . . . if you add a sum from the bottom up, and then again from the top down, the result is always different.

Mrs. La Touche – (See p. 15)

ASYMPTOTES

Frere, C.
Canning, B.
Where light *Asymptotes* o'er her bosom play,
Nor touch her glowing skin, nor intercept the day.

Quoted in Charles Edmonds
Poetry of the Anti-Jacobin
The Loves of the Triangle
Canto II, l. 122–3

BINARY

Unknown
In the binary system we count on our fists instead of on our fingers.

Source unknown

Arithmetic is where numbers fly like pigeons in and out of your head.
Carl Sandburg – (See p. 17)

BINOMIAL

Kaminsky, Kenneth
. . . yeah, our apartment was small. It was so small, we had to go out in the hall just to use the binomial expansion.

Mathematical Magazine
Professor Fogelfroe
Volume 69, Number 2, April 1996 (p. 142)

Unknown
The recorder said the fraud was of a dangerous ingenuity. "Sometimes", he said, "I am asked not to pass a sentence on a man because he is not very clever, and sometimes because he has a kink. Now I am being urged not to punish you because you suffer from a liking for permutations and combinations. I am asked to think that a man who may be a master of the binomial theorem is to be pitied more than anyone who knows nothing about it.

The Times
May 4, 1938

CALCULATION

Babbage, Charles
. . . it is said that he [Charles Babbage] sent the following letter to Alfred, Lord Tennyson about a couplet in "The Vision of Sin":

"Every minute dies a man,/Every minute one is born":

I need hardly point out to you this calculation would tend to keep the sum total of the world's population in a state of perpetual equipoise, whereas it is a well-known fact that the said sum total is constantly on the increase. I would therefore take the liberty of suggesting that in the next edition of your excellent poem the erroneous calculation to which I refer should be corrected as follows:

"Every moment dies a man/And one and a sixteenth is born."

I may add that the exact figures are 1.167, but something must, of course, be conceded to the laws of meter.

Charles Babbage and His Calculating Engines
Introduction (p. xxiii)

Belloc, Hilaire
The student must be careful in calculations involving the decimal point to put it in its exact place, neither too much to the right nor too much to the left.

The Aftermath
Appendix
(p. 147, fn)

Bennett, Charles H.
Multiplication is vexation,
  Division is as bad;
The Rule of Three it puzzles me,
  And Practice drives me mad.

Quoted in Charles H. Bennett
Old Nurse's Book of Rhymes, Jingles and Ditties (p. 25)

Berkeley, Edmund C.
The moment you have worked out an answer, start checking it—it probably isn't right.

Computers and Automation
Right Answers—A short Guide for Obtaining Them
September 1969 (p. 20)

Billings, Josh
Tew kno exackly whare the sighn iz, multiply the day ov the month bi the sighn, then find a dividend that will go into a divider four times without enny remains, subtrakt this from the sighn, add the fust quoshunt tew the last divider, then multiply the whole ov the man's boddy bi all the sighns, and the result will be jist what yu are looking after.

Old Probability: Perhaps Rain—Perhaps Not
Sighns of the Zodiac

Bloch, Felix
Erwin with his psi can do
Calculations quite a few.
But one thing has not been seen
Just what psi really mean.

Quoted in John D. Barrow
The World within the World (p. 141)

Büchner, Ludwig
Death is the surest calculation that can be made . . .

Force and Matter
Immutability of Natural Laws (p. 77)

Churchill, Winston Spencer
The human story does not always unfold like a mathematical calculation on the principle that two and two make four. Sometimes in life they make five or minus three; and sometimes the blackboard topples down in the middle of the sum and leaves the class in disorder and the pedagogue with a black eye.

Quoted in F.B. Czarnomski
The Wisdom of Winston Churchill (p. 59)
Speech
London, May 7, 1946

Dickens, Charles
. . . with affection beaming in one eye and calculation shining out of the other.

Martin Chuzzlewit
Chapter viii (p. 155)

Eisenhower, Dwight David
. . . These calculations overlook the decisive element: what counts is not necessarily the size of the dog in the fight—it's the size of the fight in the dog.

> Address to Republican National Committee
> January 31, 1958

Emerson, Ralph Waldo
Nature hates calculators.

> *The Complete Essays and other Writings of Ralph Waldo Emerson*
> Essays
> Second Series
> Experience (p. 354)

FitzGerald, Edward
For "IS" and "IS-NOT" though with Rule and Line,
And "UP-AND-DOWN" by Logic I define,
Of all that one should care to fathom, I
Was never deep in anything but—Wine.
Ah, but my Computations, People say,
Reduced the Year to better reckoning?—Nay,
'T was only striking from the Calendar
Unborn To-morrow and dead Yesterday.

> *The Rubáiyát of Omar Khayyám*
> LVI & LVII

Graham, L.A.
The professor went to the board one day
And posed to his students, just for fun,
The question, "What is the only way
To link e, i, π, zero and one?"
Up spake little Euclid, our DIAL hero,
Who handles De Moivre with infinite ease,
"Sure, $e^{i\pi} + 1 = 0$,
Give us a hard one. Next question, please!"

> *Ingenious Mathematical Problems and Methods*
> Mathematical Nursery Rhyme No. 21

Johnson, Samuel
It is wonderful when a calculation is made, how little the mind is actually employed in the discharge of any profession.

> Quoted in James Boswell
> *The Life of Samuel Johnson*
> April 6, 1775

"Nay, Madam, when you are declaiming, declaim; and when you are calculating, calculate."

Quoted in James Boswell
The Life of Samuel Johnson
April 26, 1776

Kaplan, Abraham

With such a symbolism . . . metaphysicians would no longer engage in fierce and endless controversy, but, putting their arms on each other's shoulders in friendliest fashion, they would say, "Come, let us calculate!"

The Conduct of Inquiry
Chapter V, Section 20 (p. 174)

Nietzsche, Friedrich

No more fiction for us: we calculate; but that we may calculate, we had to make fiction first.

Quoted in Tobias Dantzig
Number: The Language of Science (p. 139)

Plato

He who can properly define and divide is to be considered a god.

Quoted in Francis Bacon
Novum Organum
Second Book, 26 (near end)

I can show you that the art of computation has to do with odd and even numbers in their numerical relations to themselves and to each other.

Charmides
Section 166

Poe, Edgar Allan

. . . to calculate is not in itself to analyze.

The Complete Edgar Allan Poe Tales
The Murders in the Rue Morgue (p. 246)

Pohl, Frederik

I sat down in the back, calculating as best I could. Number forty-two. Say, at the most optimistic, an average of a minute and a half a case. That meant the judge would get to me in a little over an hour.

The Coming of the Quantum Cats
August 1983
A.M. Nicky DeSota (p. 18)

Shaw, George Bernard
And nobody can get far without at least an acquaintance with the mathematics of probability, not to the extent of making its calculations and filling examination papers with typical equations, but enough to know when they can be trusted, and when they are cooked. For when their imaginary numbers correspond to exact quantities of hard coins unalterably stamped with heads and tails, they are safe within certain limits; for here we have solid certainty ... but when the calculation is one of no constant and several very capricious variables, guesswork, personal bias, and pecuniary interests, come in so strong that those who began by ignorantly imagining that statistics cannot lie end by imagining, equally ignorantly, that they never do anything else.

> Quoted in James R. Newman
> *The World of Mathematics*
> Volume III
> The Vice of Gambling and the Virtue of Insurance (p. 1531)

Now there is no calculation that an engineer can make as to the behavior of a girder under a strain, of an astronomer as to the recurrence of a comet, more certain than the calculation that under such circumstances we shall be dismembered unnecessarily in all directions by surgeons who believe the operations to be necessary solely because they want to perform them.

> *The Doctor's Dilemma*
> Preface On Doctors (p. vi)

Thurber, James
" ... I have figured for you the distance between the horns of a dilemma, night and day, and A and Z. I have computed how far is Up, how long it takes to get Away, and what becomes of Gone. I have discovered the length of the sea serpent, the price of priceless, and the square of the hippopotamus. I know where you are when you are at Sixes and Sevens, how much Is you have to have to make an Are, and how many birds you can catch with the salt in the ocean—187,796,132, if it would interest you."

"There aren't that many birds," said the King.

"I didn't say there were," said the Royal Mathematician. "I said if there were."

> *Many Moons*

Twain, Mark
If it would take a cannon ball $3\frac{1}{3}$ seconds to travel four miles, and $3\frac{3}{8}$ seconds to travel the next four, and $3\frac{5}{8}$ to travel the next four, and if

its rate of progress continued to diminish in the same ratio, how long would it take to go fifteen hundred million miles?

Arithmeticus
Virginia, Nevada

I don't know.

Sketches Old and New
Answers to Correspondents

Unknown
As I was going to St. Ives,
I met a man with seven wives;
Every wife had seven sacks,
Every sack had seven cats,
Every cat had seven kits.
Kits, cats, sacks, and wives,
How many were going to St. Ives?

Source unknown

Wittgenstein, Ludwig
The process of *calculating* brings about just this intuition. Calculation is not an experiment.

Tractatus Logico Philosophicus
6.2331 (p. 171)

CALCULUS

Boas, Ralph P.
Rewards in Math are plenty
  But this obstacle looms big:
How can you shine in calculus
  If you won't learn any trig?

Reprinted in Ralph P. Boas, Jr.
Lion Hunting & Other Mathematical Pursuits (p. 102)

Comte, Auguste
The business of concrete mathematics is to discover the equations which express the mathematical laws of the phenomenon under consideration; and these equations are the starting point of the calculus, which must obtain from them certain quantities by means of others.

The Positive Philosophy
Volume I
Book I, Chapter II (p. 47)

Halmos, Paul
Teachers of elementary mathematics in the USA frequently complain that all calculus books are bad. That is a case in point. Calculus books are bad because there is no such subject as calculus; it is not a subject because it is many subjects. What we call calculus nowadays is the union of a dab of logic and set theory, some axiomatic theory of complete ordered fields, analytic geometry and topology, the latter in both the "general" sense (limits and continuous functions) and the algebraic sense (orientation), real-variable theory properly so called (differentiation), the combinatoric symbol manipulation called formal integration, the first steps of low-dimensional measure theory, some differential geometry, the first steps of the classical analysis of the trigonometric, exponential, and logarithmic functions, and, depending on the space available and the personal inclination of the author, some cook-book differential equations, elementary mechanics, and a small assortment of applied mathematics.

Any one of these is hard to write a good book on; the mixture is impossible.

L'Enseignement Mathématique
How to Write Mathematics
Volume 16, 1970 (p. 125)

Klein, Felix
Every one who understands the subject will agree that even the basis on which the scientific explanation of nature rests, is intelligible only to those who have learned at least the elements of the differential and integral calculus, as well as of analytical geometry.

Jahresbericht der Deutschen Mathematiker Vereinigung
Volume 11, 1902 (p. 131)

Mathematical Sciences Education Board
Although discrete mathematics and statistics provide necessary foundations for computer engineering and social sciences, calculus remains the archetype of higher mathematics. It is a powerful and elegant example of the mathematical method, leading both to major applications and to major theories. The language of calculus has spread to all scientific fields; the insight it conveys about the nature of change is something that no educated person can afford to be without.

Everybody Counts: A Report to the Nation on the Future of
Mathematics Education (pp. 51–2)

Reznick, Bruce
McEliese, Bob
Fredrickson, Hal
Hooray for calculus.
Old Newton's rootin' tootin' calculus.
The class were letting delta *x* near zero
can make a hero
of students. Teachers will say,
Just take that limit,
Be bright not dim, it's
likely to be finite and you're on your way.

Mathematical Magazine
Hooray for Calculus
Volume 61, Number 3, June 1988 (p. 147)

Thompson, Silvanus P.
Some calculus tricks are quite easy. Some are enormously difficult. The fools who write the textbooks of advanced mathematics . . . seldom take the trouble to show you how easy the easy calculations are.

Calculus Made Easy
Prologue

Tolstoy, Leo
If they'd told me at college that other people understood the integral calculus, and I didn't, then pride would have come in.

Anna Karenina
Part III
Chapter III (p. 288)

Unknown
1) You can do it by yourself and not be subjected to societal moralisms.
2) It's legal regardless of whom you do it with.
3) Only intelligent people can do it.
4) You never have to take a leak when you're through.
5) You can't get a disease from a derivative.
6) You can always be on top if you want (Product Rule) OR the bottom (Quotient Rule).
7) You never feel guilty after solving a problem.
8) You always know what your limit is.
9) All the rules are stated beforehand.
10) It lasts longer.

Why Calculus is Better Than Sex
Source unknown

O Lord, hear my anxious plea
Calculus is killing me
I know not of 'dx' or 'dy'
And probably won't until the day I die.
Please, Lord, help me in this hour
As I take my case to the highest power.
I care not for fame or loot
Just help me find one square root.
And Lord, please let me see
One passing mark in organic chemistry.
Oh such a thing I constantly dread
I'd just as soon join the Marines instead.
Lord, please give me a sign
That you've been listening all the time.
Please lead me out of this constant coma
And give me a shot at my diploma.

Source unknown

The wind blows cold,
but you don't care . . .
Blows through your soul,
Blows through your hair.
You hold a book.

It is red and white.
On the front reads Calculus,
and it fills me with fright.

Well he wears wire-rimmed glasses,
and he has white hair.
Mathematics is his game.
Pop quizzes are his snare.
If you see him in the classroom,
you just better turn and run.
'Cause it's Calculus . . .
with Mr. Lund.

If you walk into his room,
light hearted beware.
Pay attention in his class,
or do your English if you dare.
If he smiles directly at you,
and your homework isn't done,
say, "I love Calculus,"
and "Math is Fun."

Well he stands in front of class,
and he give a silly grin.
"Opportunity today,"
"Pop quiz, all right, begin."
Ya start praying to you god,
Test will take the whole day.
'Cause this is Calculus,
and he plans it that way.

Source unknown

Fill the boards with differentials,
FA-LA-LA-LA-LA-LA-LA-LA-LA
Note that du's are essential,
FA LA LA LA LA LA LA LA LA,
C's are constants here before us,
FA LA LA LA LA LA LA LA LA,
Integration cannot floor us,
FA LA LA LA LA LA LA LA LA,

Quizzes always make us queasy,
FA LA LA LA LA LA LA LA LA,
Max and mins are never easy,
FA LA LA LA LA LA LA LA LA,
Conic volumes we can measure,
FA LA LA LA LA LA LA LA LA,

Nines and tens we'll always treasure,
FA LA LA LA LA LA LA LA LA.

Fill The Boards With Differentials
(Sung to *Deck the Halls*)
Source unknown

Oh, Calculus; Oh, Calculus,
How tough are both your branches,
Oh, Calculus; Oh, Calculus,
To pass, what are my chances?
Derivatives I cannot take,
At integrals my fingers shake,
Oh, Calculus; Oh, Calculus,
How tough are both your branches.

Oh, Calculus; Oh, Calculus,
Your theorems I can't master.
Oh, Calculus; Oh, Calculus,
My proofs are a disaster.
You pull a trick out of the air,
Or find a reason, where.
Oh, Calculus; Oh, Calculus,
Your theorems I can't master.

Oh Calculus, Oh Calculus!
(Sung to *Oh Christmas Tree*)
Source unknown

CIRCLE

Browne, Sir Thomas
Circles and right lines limit and close all bodies, and the mortal right-lined circle must conclude and shut up all.

Hydriotaphia
Chapter V

Chesterton, Gilbert Keith
A small circle is quite as infinite as a large circle.

Orthodoxy
The Maniac (p. 33)

Emerson, Ralph Waldo
The eye is the first circle; the horizon which it forms is the second; and throughout nature this primary figure is repeated without end. It is the highest emblem in the cipher of the world.

The Complete Essays and other Writings of Ralph Waldo Emerson
Essays
Circles (p. 279)

Pope, Alexander
As the small pebble stirs the peaceful lake;
The centre mov'd, a circle straight succeeds,
Another still, and still another spreads.

The Complete Poetical Works
Essays on Man
Epistle IV, l. 364

Shaw, George Bernard
KNELLER: Just what such blockheads would believe. The circle is a dead thing like a straight line: no living hand can draw it: you make it by twirling a pair of dividers.

The Complete Plays of Bernard Shaw
In Good King Charles's Golden Days
Act I (p. 1358)

Valentinus, Jacobus Falco
At first a circle I was called,
And was a curve around about
Like lofty orbit of the sun
Or rainbow arch among the clouds.
A noble figure then was I—
And lacking nothing but a start,
And lacking nothing but an end.

Quoted in Augustus de Morgan
A Budget of Paradoxes
Volume I (p. 54, fn)

Death is the surest calculation that can be made . . .
Ludwig Büchner – (See p.24)

CLASS

Whitehead, Alfred North
But in the prevalent discussion of classes, there are illegitimate transitions to the notions of a 'nexus' and of a 'proposition'. The appeal to a class to perform the services of a proper entity is exactly analogous to an appeal to an imaginary terrier to kill a real rat.

Process and Reality
An Essay in Cosmology
The Theory of Feelings (p. 348)

COMMON SENSE

Arnauld, Antoine
Common sense is not really so common.

The Art of Thinking: Port-Royal Logic
First Discourse (p. 9)

Bell, Eric T.
This is precisely what common sense is for, to be jarred into uncommon sense. One of the chief services which mathematics has rendered the human race in the past century is to put 'common sense' where it belongs, on the topmost shelf next to the dusty canister labeled 'discarded nonsense.'

Mathematics: Queen and Servant of Science
Mathematical Truth (pp. 17–18)

Einstein, Albert
. . . common sense is nothing more than a deposit of prejudices laid down in the mind before you reach eighteen.

Quoted in Eric T. Bell
Mathematics: Queen and Servant of Science
Breaking Bounds (p. 42)

Holmes, Oliver Wendell
Science is a first-rate piece of furniture for a man's upper chamber, if he has common sense on the ground floor.

The Poet at the Breakfast-Table
Chapter V (p. 140)

Oppenheimer, Julius Robert
. . . distrust all the philosophers who claim that by examining science they come to the results in contradiction with common sense. Science is based on common sense; it cannot contradict it.

In University of Denver
Foundations for World Order
The Scientific Foundations for World Order (p. 51)

Russell, Bertrand

Common sense starts with the notion that there is matter where we can get sensations of touch, but not elsewhere. Then it gets puzzled by wind, breath, clouds, etc., whence it is led to the conception of "spirit"—I speak etymologically. After "spirit" has been replaced by "gas," there is a further stage, that of the aether.

The Analysis of Matter
Chapter XIII (p. 121)

The supposition of common sense and naive realism, that we see the actual physical object, is very hard to reconcile with the scientific view that our perception occurs somewhat later than the emission of light by the object; and this difficulty is not overcome by the fact that the time involved, like the notorious baby, is a very little one.

The Analysis of Matter
Chapter XV (p. 155)

Common sense, however it tries, cannot avoid being surprised from time to time. The aim of science is to save it from such surprises.

Quoted in Jean-Pierre Luminet
Black Holes (p. 182)

Titchener, E.B.

. . . common sense is the very antipodes of science.

Systematic Psychology
Chapter I (p. 48)

Thomson, William (Lord Kelvin)

Do not imagine . . . that mathematics is hard and crabbed, and repulsive to common sense. It is merely the etherealization of common sense.

In S.P. Thompson
The Life of William Thomson Baron Kelvin of Largs
Volume II
Views and Opinions (p. 1139)

Whitehead, Alfred North

Now in creative thought common sense is a bad master. Its sole criterion for judgment is that the new ideas shall look like the old ones. In other words it can only act by suppressing originality.

An Introduction to Mathematics
Chapter 11 (p. 116)

COMPUTERS

Klein, Felix
If the activity of a science can be supplied by a machine, that science cannot amount to much, so it is said; and hence it deserves a subordinate place. The answer to such arguments, however, is that the mathematician, even when he is himself operating with numbers and formulas, is by no means an inferior counterpart of the errorless machine, "thoughtless thinker" of Thomas; but rather, he sets for himself his problems with definite, interesting, and valuable ends in view, and carries them to solution in appropriate and original manner. He turns over to the machine only certain operations which recur frequently in the same way, and it is precisely the mathematician—one must not forget this—who invented the machine for his own relief, and who, for his own intelligent ends, designates the tasks which it shall perform.

Elementary Mathematics from an Advanced Standpoint (p. 22)

Mathematical Sciences Education Board
Calculators and computers should be used in ways that anticipate continuing rapid change due to technological developments. Technology should be used not because it is seductive, but because it can enhance mathematical learning by extending each student's mathematical power.

Everybody Counts: A Report to the Nation on the Future of
Mathematics Education (p. 84)

CONIC SECTIONS

Whewell, William
If the Greeks had not cultivated Conic Sections, Kepler could not have superseded Ptolemy; if the Greeks had cultivated Dynamics, Kepler might have anticipated Newton.

History of the Inductive Science
Volume I (p. 311)

CURVES

Kasner, Edward
Newman, James R.
The curves treated by the calculus are normal and healthy; they possess no idiosyncrasies. But mathematicians would not be happy merely with simple, lusty configurations. Beyond these their curiosity extends to psychopathic patients, each of whom has an individual case history resembling no other; these are the pathological curves in mathematics.

Mathematics and the Imagination (p. 343)

Klein, Felix
Everyone knows what a curve is, until he has studied enough mathematics to become confused through the countless number of possible exceptions.

Quoted in Carl B. Boyer
Scientific American
The Invention of Analytic Geometry
Volume 180, Number 1, January 1949 (p. 41)

West, Mae
A figure with curves always offers a lot of interesting angles.

The Wit and Wisdom of Mae West (p. 35)

DECIMALS

Churchill, Lord Randolph
. . . I could never make out what those damned dots meant.

Quoted in Winston S. Churchill
Lord Randolph Churchill
Volume II
Chapter XV (p. 184)

Unknown
Decimals have a point.

Source unknown

DEADUCTION

Descartes, René
... the two operations of our understanding, intuition and deduction, on which alone we have said we must rely in the acquisition of knowledge.

Rules for the Direction of the Mind
Rule IX

Jevons, W. Stanley
Deduction is certain and infallible, in the sense that each step in deductive reasoning will lead us to some result, as certain as the law itself. But it does not follow that deduction will lead the reasoner to every result of a law or combination of laws.

The Principles of Science (p. 534)

Reid, Thomas
The mathematician pays not the least regard either to testimony or conjecture, but deduces everything by demonstrative reasoning, from his definitions and axioms. Indeed, whatever is built upon conjecture, is improperly called science; for conjecture may beget opinion, but cannot produce knowledge.

Essays on the Intellectual Powers of Man
Essay 1, Chapter 3

Unknown
Deductive Process:
Formulate hypothesis
Apply for grant
Perform experiments or gather data to test hypothesis
Revise hypothesis to fit data
Backdate revised hypothesis
Publish.

Source unknown

Whewell, William
These sciences have no principles besides definitions and axioms, and no process of proof but *deduction*; this process, however, assuming a most remarkable character; and exhibiting a combination of simplicity and complexity, of rigour and generality, quite unparalleled in other subjects.

The Philosophy of the Inductive Sciences
Volume I
Part 1, Book 2, Chapter 1, section 2 (p. 83)

Whitehead, Alfred North
Mathematical reasoning is deductive in the sense that it is based upon definitions which, as far as the validity of the reasoning is concerned (apart from any existential import) needs only the test of self-consistency. Thus no external verification of definitions is required in mathematics, as long as it is considered merely as mathematics.

A Treatise on Universal Algebra
Preface (p. vi)

DEFINED

de Morgan, Augustus

. . . there are terms which cannot be defined, such as number and quantity. Any attempt at a definition would only throw difficulty in the student's way, which is already done in geometry by the attempts at an explanation of the terms point, straight line, and others, which are to be found in treatises on that subject. A point is defined to be that "which has no parts and which has no magnitude"; a straight line is that which "lies evenly between its extreme points." . . . In this case the explanation is a great deal harder than the term to be explained, which must always happen when we are guilty of the absurdity of attempting to make the simplest ideas yet more simple.

On the Study and Difficulties of Mathematics (pp. 12–13)

DERIVATIVE

Berkeley, George
But he who can digest a second or third Fluxion, a second or third Difference, need not, methinks, be squeamish about any Point in Divinity.

The Analyst
Subsection 7 (p. 168)

And what are these Fluxions? The Velocities of evanescent Increments. And what are these same evanescent Increments? They are neither finite Quantities, nor Quantities infinitely small, nor yet nothing. May we not call them Ghosts of departed Quantities?

The Analyst
Section 35 (p. 199)

Lehrer, Tom
You take a function of x and you call it y,
Take any x-nought that you care to try,
You make a little change and call it delta x,
The corresponding change in y is what you find nex',
And then you take the quotient and now carefully
Send delta x to zero, and I think you'll see
That what the limit gives us, if our work all checks,
Is what we call dy/dx,
It's just dy/dx.

American Mathematical Monthly
The Derivative Song
Volume 81, Number 5, May 1974 (p. 490)

DIFFERENTIAL

Zamyatin, Yevgeny
The function of man's highest faculty, his reason, consists precisely of the continuous limitation of infinity, the breaking up of infinity into convenient, easily digestible portions—differentials.

We
Twelfth Entry (p. 58)

DIFFERENTIAL EQUATION

Born, Max
If God has made the world a perfect mechanism, He has at least conceded so much to our imperfect intellect that in order to predict little parts of it, we need not solve innumerable differential equations, but can use dice with fair success.

Quoted in Heinz R. Pagels
The Cosmic Code (p. 73)

The difficulty involved is that the proper and adequate means of describing changes in continuous deformable bodies is the method of differential equations . . .. They express mathematically the physical concept of contiguous action.

Einstein's Theory of Relativity
Chapter IV, Section 6 (p. 109 & p. 111)

Einstein, Albert
Thus the partial differential equation entered theoretical physics as a handmaid, but has gradually become mistress.

The World As I See It (p. 63)

Fourier, [Jean Baptiste] Joseph
The differential equations of the propagation of heat express the most general conditions, and reduce the physical questions to problems of pure analysis, and this is the proper object of theory.

Analytical Theory of Heat
Preliminary Discourse

Poincaré, Henri
If one looks at the different problems of the integral calculus which arise naturally when he wishes to go deep into the different parts of physics, it is impossible not to be struck by the analogies existing. Whether it be electrostatics or electrodynamics, the propagation of heat, optics,

elasticity, or hydrodynamics, we are led always to differential equations of the same family.

American Journal of Mathematics
Sur les Équations aux Dérivées Partielles de la Physique Mathématique
Volume 12, 1890 (p. 211)

Pólya, George
In order to solve a differential equation you look at it till a solution occurs to you.

How to Solve It (p. 181)

Sholander, Marlow
If the finding is essential
　Of a singular solution
For equations differential,
　Let me sketch its execution.
First obtain some ordinary
　Members of the family
Of solutions—oh, not very
　Many, maybe twenty three.
Find their curves by carefully plotting.
　Ink them quickly and you ought,
From the points requiring blotting,
　To obtain the locus sought.

Mathematics Magazine
Envelopes and Nodes
Volume 34, Number 2, November–December 1960 (p. 108)

Turing, Alan
Science is a differential equation. Religion is a boundary condition.

Quoted in John D. Barrow
Theories of Everything (p. 31)

Unknown
Trying to solve differential equations is a youthful activity that you will soon grow tired of.

Source unknown

There once was a man from Lapeze
Who spent his life solving d.e.'s.
　"They're easy", he said,
　"But partials I dread,
I break out in spots and I wheeze!"

Source unknown

There once was a letter called tau
But no-one could figure out how
    Some people were sure
    It was pronounced 'tor',
But arguments still persist now.

Source unknown

A man had a certain evasion
For solving all difference equations.
    He used random numbers
    To cover his blunders,
And answers caused quite a sensation!

Source unknown

A man I once knew was a tutor
Who got himself into a stupor.
    When d.e.'s were found
    With errors unbound,
He put his fist through his computer!

Source unknown

A subject we did had a facility
For testing equations' stability.
    When things did not work,
    The class went berserk,
And erupted in violent hostility.

Source unknown

The whole point of mathematics is to solve differential equations!
Mathematical and Scientific Quotes From Cambridge
The Internet

Nature abhors second order differential equations.
Mathematical and Scientific Quotes From Cambridge
The Internet

Whitehead, Alfred North
Matter-of-fact is an abstraction, arrived at by confining thought to purely formal relations which then masquerade as the final reality. This is why science, in its perfection, relapses into the study of differential equations. The concrete world has slipped through the meshes of the scientific net.
Modes of Thought
Creative Impulse
Importance (p. 25)

DIMENSION

Abbott, Edwin A.
Yet I exist in the hope that these memoirs . . . may find their way to the minds of humanity in Some Dimension, and may stir up a race of rebels who shall refuse to be confined to limited Dimensionality.

Flatland
Part II, section 22 (p. 107)

Baker, W.R.
Length, breadth, and depth are said to be
The limits of man's comprehension,
    But when I see the pile of junk
    That she can get into a trunk
The mystery convinces me
That woman knows a fourth dimension.

Harper's Magazine
The Magic Box
Volume CLVI, December 1927–May 1928 (p. 649)

Eddington, Sir Arthur Stanley
The quest of the absolute leads into the four-dimensional world.

The Nature of the Physical World (p. 26)

Fock, Vladimir Alexandrovich
Though we may weigh it as we will,
Exhausted and delirious,
One-hundred-thirty-seven still
Remains for us mysterious.
But Eddington, *he*, sees it clear,
Denouncing those who tend to jeer;
It is the number of (says he)
The world's dimensions. *Can it ?!be?!*—

Quoted in George Gamow
Biography of Physics (p. 327)

51

Frost, Robert
GOD. That's about right.
I should have said. You got your age reversed
When time was found to be a space dimension.
That could, like any space, be turned around in?

The Poetry of Robert Frost
A Masque of Reason
l. 161

Hamilton, William Rowan
Time is said to have only *one dimension*, and space to have *three dimensions*
. . .. The mathematical *quaternion* partakes of *both* these elements; in
technical language it may be said to be "time plus space", or "space
plus time": and in this sense it has, or at least involves a reference to,
four dimensions . . .

And how the One of Time, of Space the Three,
Might in the Chain of Symbols girdled be.

Quoted in Robert Percevel Graves
Life of Sir William Rowan Hamilton
Volume 3 (p. 635)

Maxwell, James Clerk
My soul is an entangled knot,
Upon a liquid vortex wrought
By Intellect in the Unseen residing,
And thine doth like a convict sit,
With marlinespike untwisting it,
Only to find its knottiness abiding;
Since all the tools for its untying
In four-dimensional space are lying,
Wherein thy fancy intersperses
Long avenues of universes,
While Klein and Clifford fill the void
With one finite, unbounded homaloid,
And think the Infinite is now at last destroyed.

The Life of James Clerk Maxwell
A Paradoxical Ode (p. 649)

Reichenbach, Hans
Let us assume that the three dimensions of space are visualized in the
customary fashion, and let us substitute a color for the fourth dimension.
Every physical object is liable to changes in color as well as in position.
An object might, for example, be capable of going through all shades
from red through violet to blue. A physical interaction between any two
bodies is possible only if they are close to each other in space as well as

in color. Bodies of different colors would penetrate each other without interference . . . If we lock a number of flies into a red glass globe, they may yet escape: they may change their color to blue and then able to penetrate the red globe.

The Philosophy of Space & Time
Section 44 (pp. 281–2)

Updike, John

. . . you need no more or less than three dimensions to make a knot, a knot that tightens on itself and won't pull apart, and that's what the ultimate particles are—knots in space-time. You can't make a knot in two dimensions because there's no over or under . . .

Roger's Version (p. 302)

Imagine nothing, a total vacuum. But wait! There's something in it! Points, potential geometry. A kind of dust of structureless points. Or, if that's too woolly for you, try 'a Borel set of points not yet assembled into a manifold of any particular dimensionality.'

Roger's Version (p. 303)

Uspenskii, Petr Demianovich

And when we shall see or feel ourselves in the world of four dimensions we shall see that the world of three dimensions does not really exist and has never existed: that it was the creation of our own fantasy, a phantom host, an optical illusion, a delusion—anything one pleases excepting only reality.

Tertium Organum
Chapter IX (p. 98)

Wells, H.G.

There are really four dimensions, three of which we call the three planes of Space, and a fourth, Time. There is, however, a tendency to draw an unreal distinction between the former three dimensions and the latter, because it happens that our consciousness moves intermittently in one direction along the latter from the beginning to the end of our lives.

The Short Stories of H.G. Wells
The Time Machine (p. 4)

Mathematical theorists tell us that the only way in which the right and left sides of a solid body can be changed is by taking that body clean out of space as we know it—taking it out of ordinary existence, that is and turning it somewhere outside space . . . To put the thing in technical language, the curious inversion of Plattner's right and left sides is proof that he has moved out of our space into what is called the Fourth Dimension, and that he has returned again to our world.

The Short Stories of H.G. Wells
The Plattner Story (p. 329)

Whitehead, Alfred North
I regret that it has been necessary for me in this lecture to administer a large dose of four-dimensional geometry. I do not apologise, because I am really not responsible for the fact that nature in its most fundamental aspect is four-dimensional. Things are what they are . . .

The Concept of Nature
Space and Motion (p. 119)

Williams, W.H.
And space, it has dimensions four,
  Instead of only three.
The square of the hypotenuse
  Ain't what it used to be.

Quoted in Fred Alan Wolf
Parallel Universes (p. 105)

Length, breadth, and depth are said to be
The limits of man's comprehension,
But when I see the pile of junk
That she can get into a trunk
The mystery convinces me
That woman knows a fourth dimension.
W.R. Baker – (See p. 51)

DISCOVERY

Bruner, Jerome Seymore
First, I should be clear about what the act of discovery entails. It is rarely, on the frontier of knowledge or elsewhere, that new facts are "discovered" in the sense of being encountered, as Newton suggested, in the form of islands of truth in an uncharted sea of ignorance. Or if they appear to be discovered in this way, it is almost always thanks to some happy hypothesis about where to navigate. Discovery, like surprise, favors the well-prepared mind.

On Knowing—Essays for the Left Hand
Part I
The Quest for Clarity (p. 82)

Leibniz, Gottfried Wilhelm
It is an extremely useful thing to have knowledge of the true origins of memorable discoveries, especially those that have been found not by accident but by dint of meditation. It is not so much that thereby history may attribute to each man his own discoveries and that others should be encouraged to earn like commendation, as that the art of making discoveries should be extended by considering noteworthy examples of it.

Quoted in J.M. Child
The Early Mathematical Manuscripts of Leibniz (p. 22)

Pólya, George
A GREAT discovery solves a great problem but there is a grain of discovery in the solution of any problem.

How to Solve It (p. v)

The first rule of discovery is to have brains and good luck. The second rule of discovery is to sit tight and wait till you get a bright idea.

How to Solve It (p. 158)

Szent-Györgyi, Albert
Discovery consists of seeing what everybody has seen and thinking what nobody has thought.

Quoted in Irving John Good (Editor)
The Scientist Speculates (p. 15)

Whitehead, Alfred North
What Bacon omitted was the play of free imagination, controlled by the requirements of coherence and logic. The true method of discovery is like the flight of an aeroplane. It starts from the ground of particular observation; it makes a flight in the thin air of imaginative generalization; and it again lands for renewed observation rendered acute by rational interpretation.

Process and Reality
An Essay in Cosmology
Speculative Philosophy (p. 7)

Can you do Division? Divide a loaf by a knife—what's the answer to *that*?

Lewis Carroll – (See p. 58)

DIVERGENCE

Lillich, Robert
Divergence B, it's plain to see, is zero.
   I think they've got it, I think they've got it!
And del dot D is always rho, you know.
   I think they've got it, I think they've got it!

The Physics Teacher
My Fair Physicist
December 1968 (p. 490)

DIVISION

Adams, Douglas

It is known that there is an infinite number of worlds, but that not every one is inhabited. Therefore, there must be a finite number of inhabited worlds. Any finite number divided by infinity is as near to nothing as makes no odds, so if every planet in the Universe has a population of zero then the entire population of the Universe must also be zero, and any people you may actually meet from time to time are merely the products of a deranged imagination.

The Original Hitchhiker Radio Script
Fit the Fifth (p. 102)

Carroll, Lewis

Can you do Division? Divide a loaf by a knife—what's the answer to *that?*

The Complete Works of Lewis Carroll
Through the Looking Glass
Chapter IX

Kemble, William H.

Allow me to introduce you to my particular friend Mr. George O Evans ... He understands Addition, Division, and Silence.

New York Sun
20 June 1872

e

Berrett, Wayne
It had to be e,
nonintegral e,
I looked around
Until I found
A base that would do.
To diff'rentiate
or to integrate,
One that would not
Carry along
Some ugly weight.

Some bases I know
Are Simpler to state,
A snap to invert,
Exponentiate,
But they wouldn't do.
For no other base can fit math so well,
With all its digits I love it still!
It had to be e,
Irrational e,
It had to be e!

Mathematics Magazine
It Had to be e
Volume 68, Number 1, February 1995 (p. 15)

Brewster, G.W.
$2^{(5/2)\wedge 2/5} = e$

The Mathematical Gazette
Volume 25, Number 263, February 1941 (p. 49)

Klein, Felix

The definition of e is usually, in imitation of the French models, placed at the very beginning of the great text books of analysis, and entirely unmotivated, whereby the really valuable element is missed, the one which mediates the understanding, namely, an explanation of why precisely this remarkable limit is used as base and why the resulting logarithms are called natural.

Elementary Mathematics from an Advanced Standpoint (p. 146)

Teller, Edward

. . . the first nine digits after the decimal can be remembered by e = 2.7(Andrew Jackson)^2, or e = 2.718281828 . . ., because Andrew Jackson was elected President of the United States in 1828. For those good in mathematics on the other hand, this is a good way to remember their American History.

Conversations on the Dark Secrets of Physics (p. 87)

ELLIPSE

Bell, Eric T.
A circle no doubt has a certain appealing simplicity at the first glance, but one look at a healthy ellipse should have convinced even the most mystical of astronomers that that the perfect simplicity of the circle is akin to the vacant smile of complete idiocy. Compared to what an ellipse can tell us, a circle has nothing to say.

The Handmaiden of the Sciences (p. 26)

ELLIPTIC FUNCTIONS

Bellman, Richard
The theory of elliptic functions is the fairyland of mathematics. The mathematician who once gazes upon this enchanting and wondrous domain crowded with the most beautiful relations and concepts is forever captivated.

A Brief Introduction to Theta Functions (p. vii)

EQUATION

Dirac, Paul Adrien Maurice
I consider that I understand an equation when I can predict the properties of its solutions, without actually solving it.

Quoted in Frank Wilczek and Betsy Devine
Longing for the Harmonies (p. 102)

. . . it is more important to have beauty in one's equations than to have them fit experiment.

Scientific American
The Evolution of the Physicist's Picture of Nature
Volume 208, Number 5, May 1963 (p. 47)

If one is working from the point of view of getting beauty in one's equation, . . . one is on a sure line of progress.

Scientific American
The Evolution of the Physicist's Picture of Nature
Volume 208, Number 5, May 1963 (p. 47)

Feynman, Richard P.
Where did we get that [Schrödinger's equation] from? It's not possible to derive it from anything you know. It came out of the mind of Schrödinger
. . .

The Feynman Lectures on Physics
Volume III
Chapter 16-5 (p. 16-12)

Hawking, Stephen
Even if there is only one possible unified theory, it is just a set of rules and equations.

A Brief History of Time (p. 174)

Hertz, Heinrich
Maxwell's theory is Maxwell's system of equations.

Electric Waves; Being Researches on the Propagation of
Electric Action with Finite Velocity through Space (p. 21)

Jackins, Harvey
If you don't read poetry how the hell can you solve equations?

Quoted in Julian Weissglass
Exploring Elementary Mathematics

Thoreau, Henry David
I do believe in simplicity. When the mathematician would solve a difficult problem he first frees the equation from all encumbrances, and reduces it to its simplest terms.

Quoted in William Peterfield Trent
Cambridge History of American Literature
Book II (Continued)
Chapter X
Thoreau (p. 8)

DOC - YOU'VE GOT TO HELP ME - IT'S THOSE DAMNED DOTS..!

. . . I could never make out what those damned dots meant.
Lord Randolph Churchill – (See p. 42)

ERROR

Adams, Franklin Pierce
If frequently I fret and fume,
And absolutely will not smile,
I err in company with Hume,
Old Socrates and T. Carlyle.

Tobogganing on Parnassus
Erring in Company

Beard, George M.
As quantitative truth is of all forms of truth the most absolute and satisfying, so quantitative error is of all forms of error the most complete and illusory.

Popular Science Monthly
Experiments with Living Human Beings (p. 751)
Volume 14, April, 1879

da Vinci, Leonardo
O mathematicians, throw light on this error.

The Notebooks of Leonardo da Vinci
Volume I
Philosophy (p. 70)

Diamond, Solomon
Error does not carry any recognizable badge, for when we change our point of view, to focus on a different problem, what had been error may become information, and what had been information may become error.

Information and Error (p. 7)

Here, by the grace of Chance, we've staked a Mean,
Uncertain marker of elusive Truth.
But have we caught a fact, or trapped a doubt
Within this stretching span of confidence—
A shadow world four standard errors wide,

All swollen by the stint of observation?
For recollect that once in twenty times
The phantom Truth will even lie beyond
That span, in the unending thin-drawn tails
Which point to the infinitude of Error.

Information and Error (p. 120)

Heinlein, Robert A.
I shot an error into the air. It's still going . . . *everywhere.*

Expanded Universe (p. 514)

Newton, Sir Isaac
. . . the errors are not the art, but in the artifiers.

Mathematical Principles of Natural Philosophy
Preface to the First Edition

Siegel, Eli
When truth is divided, errors multiply.

Damned Welcome
Aesthetic Realism Maxims
Part II, #360 (p. 147)

Tupper, Martin F.
Error is a hardy plant; it flourishes in every soil.

Proverbial Philosophy: A Book of Thoughts and Arguments
Of Truth in Things False (p. 5)

EUCLID

Dostoevsky, Fyodor Mikhailovich
. . . if God exists and if He really did create the world, then, as we all know, He created it according to the geometry of Euclid and the human mind with the conception of only three dimensions in space. Yet there have been and still are geometricians and philosophers, and even some of the most distinguished, who doubt whether the whole universe, or to speak more widely, the whole of being, was only created in Euclid's geometry; they even dare to dream that two parallel lines, which according to Euclid can never meet on earth, may meet somewhere in infinity. I have come to the conclusion that, since I can't understand even that, I can't expect to understand about God.

The Brothers Karamozov
Book V, Chapter 3

Keyser, Cassius J.
The Elements of Euclid is as small a part of mathematics as the Iliad is of literature; or as the sculpture of Phidias is of the world's total art.

Quoted in Columbia University
Lectures on Science, Philosophy and Art 1907–1908 (p. 8)

Lindsay, Vachel
Old Euclid drew a circle
On a sand-beach long ago,
He bound it and enclosed it
With angles thus and so.

The Congo and Other Poems
Euclid

Turner, H.H.
When Euclid framed his definitions
  He did not miss "the point";
Space was prescribed by his conditions
  For angles twain conjoint.

The Mathematical Gazette
Volume VI, Number 100, October 1912 (p. 403)

EXAMPLE

Halmos, Paul R.
A good stock of examples, as large as possible, is indispensable for a thorough understanding of any concept, and when I want to learn something new, I make it my first job to build one.

Quoted in Joseph A. Gallian
Contemporary Abstract Algebra (p. 34)

FACTORS

Milne, A.A.
Suddenly Christopher Robin began to tell Pooh about some of the things:
People called Kings and Queens and something called Factors . . .

The House At Pooh Corner (p. 171)

. . . and then, as Pooh seemed disappointed, he added quickly, "but it's
grander than Factors."

The House At Pooh Corner (p. 175)

FIBONACCI

Baumel, Judith
Learn the particular strength
of the Fibonacci series,
a balanced spiraling
outward of shapes,
those golden numbers
which describe dimensions
of sea shells, rams' horns,
collections of petals
and generations of bees.

The Weight of Numbers
Fibonacci (p. 21)

Lindon, J.A.
Each wife of Fibonacci,
Eating nothing that wasn't starchy
Weighed as much as the two before her.
His fifth was some signora!

Quoted in Martin Gardner
Mathematical Circus (p. 152)

Unknown
Fibonacci is not a shortened form of the Italian name that is actually
spelled: F i bb ooo nnnnn aaaaaaaa cccccccccccccccccccccccccccccccccccccc
iiiiiiiiiiiiiiiiiiiiiiiiiiiiiiiiiiiiii.

Source unknown

FIELD

Dantzig, Tobias

Today we know that possibility and impossibility have each only a relative meaning; that neither is an intrinsic property of the operation but merely a *restriction which human tradition has imposed on the field of the operand*. Remove the barrier, extend the field, and the impossible becomes possible.

Number: The Language of Science (p. 89)

FIGURES

Carlyle, Thomas
A witty statesman said you might prove anything by figures.

English and other Critical Essays
Chartism
Chapter II (p. 170)

Crichton, Michael
"John. Trust us on this, we have the figures. We are telling you with ninety-five percent confidence intervals how the people feel."

Rising Sun (p. 255)

de Saint-Exupéry, Antoine
Grown-ups love figures. When you tell them that you have made a new friend, they never ask you any questions about essential matters. They never say to you, "What does his voice sound like? What games does he love best? Does he collect butterflies?" Instead, they demand: "How old is he? How many brothers has he? How much does he weigh? How much money does his father make?" Only from these figures do they think they have learned anything about him.

The Little Prince (p. 16)

Goethe, Johann Wolfgang von
Man hat behauptet, die Welt werde durch Zahlen regiert: das aber weiss ich, dass die Zahlen uns belehren, ob sie gut oder schlecht regiert werde.
[It has been said that figures rule the world; maybe. I am quite sure that it is figures which show us whether it is being ruled well or badly.]

Conversation with Eckerman, 31st January 1830

Hopkins, Harry
Figures are faceless and incestuous.

The Numbers Game: the Bland Totalitarianism (p. 15)

Huxley, Aldous
"Give them a few figures, Mr. Foster," said the Director, who was tired of talking.

Brave New World (p. 11)

Proverb, English
Figures never lie.

Sage, M.
. . . battalions of figures are like battalions of men, not always as strong as is supposed.

Mrs. Piper and the Society for Psychical Research
Chapter XV (p. 151)

Tarbell, Ida
There is no more effective medicine to apply to feverish public sentiment than figures. To be sure, they must be properly proposed, must cover the case, must confine themselves to a quarter of it, and they must be gathered for their own sake, not for the sake of a theory.

The Ways of Woman (p. 3)

Unknown
In any collection of data, the figure that is most obviously correct—beyond all need of checking—is the mistake.

Source unknown

FORMULA

Dudley, Underwood
Formulas should be useful. If not they should be astounding, elegant, enlightening, simple, or have some other redeeming value.

Mathematics Magazine
Formulas for Primes (p. 22)
Volume 56, Number 1, January 1983

Authors who discover formulas should not rush into print. Even as in business and marriage, in mathematics not *all* that is true needs to be published.

Mathematics Magazine
Formulas for Primes (p. 22)
Volume 56, Number 1, January 1983

Kasner, Edward
Newman, James R.
There is a famous formula,—perhaps the most compact and famous of all formulas—developed by Euler from a discovery of the French mathematician De Moivre: $e^{i\pi} + 1 = 0$ . . . It appeals equally to the mystic, the scientist, the philosopher, the mathematician.

Mathematics and the Imagination (p. 103)

Kipling, Rudyard
No proposition Euclid wrote
  No formulae the text-books know,
Will turn the bullet from your coat,
  Or ward the tulwar's downward blow.
Strike hard who cares—shoot straight who can—
The odds are on the cheaper man.

Rudyard Kipling's Verse: Inclusive Edition
Arithmetic on the Frontier (p. 45)

Peirce, Charles Sanders

It is terrible to see how a single unclear idea, a single formula without meaning, lurking in a young man's head, will sometimes act like an obstruction of inert matter in an artery, hindering the nutrition of the brain, and condemning its victim to pine away in the fullness of his intellectual vigor and in the midst of intellectual plenty.

Chance, Love and Logic
Part I, Second Paper (p. 37)

Unknown

fo^r[M−u/l]a

Source unknown

THAT'S A MIGHTY FINE
EUCLIDEAN KNOT
THERE PARTNER ...!

The cowboys have a way of trussing up a steer or a pugnacious bronco which fixes the brute so that it can neither move nor think. This is the hog-tie, and it is what Euclid did to geometry.

Eric T. Bell – (See p. 81)

FRACTIONS

Burr, Lehigh
"My daughter," and his voice was stern,
  "You must set this matter right;
What time did the Sophomore leave,
  Who sent in his card last night?"
"His work was pressing, father dear,
  And his love for it was great;
He took his leave and went away
  Before a quarter of eight."
Then a twinkle came to her bright blue eye,
  And her dimple deeper grew.
"Tis surely no sin to tell him that,
  For a quarter of eight is two."

Quoted in R.L. Paget
Poetry of American Wit and Humor
Applied Mathematics (p. 302)

Fosdick, Harry Emerson
. . . when life ceases to be a fraction and becomes an integer.

On Being A Real Person (p. 33)

FUNCTION

Gauss, Carl Friedrich
One should not forget that the functions, like all mathematical constructions, are only our own creation, and that when the definition with which one begins ceases to make sense, one should not ask, what is, but what is convenient to assume in order that it remains significant.

Quoted in Bryan H. Bunch
Mathematical Fallacies and Paradoxes (p. 1)

Keyser, Cassius J.
Every one is familiar with the *ordinary* notion of a function—with the notion, that is, of the lawful dependence of one or more variable things upon other variable things, as the area of a rectangle upon the lengths of its sides, as the distance traveled upon the rate of going, as the volume of a gas upon the temperature and pressure, as the prosperity of a throat specialist upon the moisture of the climate, as the attraction of material particles upon the distance asunder, as the rate of chemical change upon the amount or the mass of the substance involved, as the turbulence of labor upon the lust of capital, and so on and on without end.

Mathematical Philosophy
Lecture III (pp. 49–50)

The connections shown by these particular examples hold in general: given a transformation, you have a function and a relation; given a function, you have a relation and a transformation; given a relation, you have a transformation and a function: *one* thing—*three* aspects; and the fact is exceedingly interesting and weighty.

Mathematical Philosophy
Lecture X (pp. 167–8)

McCormack, Thomas J.
. . . in that flower of modern mathematical thought—the notion of a function.

The Monist
On the Nature of Scientific Law and Scientific Explanation
Volume 10, Number 4, July 1900 (p. 555)

Rankine, W.J.M.
Let x denote beauty, y manners well-bred,
z fortune (this last is essential),
Let L stand for love—our philosopher said—
Then L is a function of x, y and z
Of the kind that is known as potential.

> Quoted in Sir Arthur Eddington
> *The Philosophy of Physical Science*
> Chapter IX (p. 138)
> *Songs and Fables*
> The Mathematician in Love
> Verse 6

Roe, E.D., Jr.
The continuous function is the only workable and usable function. It alone is subject to law and the laws of calculation. It is a loyal subject of the mathematical kingdom. Other so-called or miscalled functions are freaks, anarchists, disturbers of the peace, malformed curiosities which one and all are of no use to anyone, least of all to the loyal and burden-bearing subjects who by keeping the laws maintain the kingdom and make its advance possible.

. . . scholarship lies in the direction of paying deference to the loyal continuous function rather than to the outlaws of mathematical society.

> *The Mathematics Teacher*
> A Generalized Definition of Limit
> Volume III, Number 1, September 1910 (p. 4)

Unknown
The famous professor of mathematics was in town for a conference. Since he had some free time, he was approached to give a seminar for the undergraduate mathematics students at the local college.

After covering several blackboards with densely packed computations and expressions filled with Bessel functions and more, the professor remembered that there were many undergraduate students in the room. Feeling just a twinge of remorse that perhaps he was talking above the heads of some of the students in his audience, he turned around and asked the audience if there were any students who had never seen a Bessel function. The audience was silent for a moment. Finally, one intrepid student raised his hand to admit that he had never seen Bessel functions. The professor nodded with apparent comprehension. Without hesitation, he turned around and pointed at the blackboard, while saying "well, there's one now" and continued his talk.

> Source unknown

GEOMETER

Du Hamel, Joannes Baptiste
I do not find that geometers are mighty solicitous, whether their arguments be, in formula, compounded according to logical prescription; and yet there are none who demonstrate wither more precisely or with greater conviction. For they usually follow the guidance of nature; descending step by step, from the simpler and more general to the more complex, and defining every term, they leave no ambiguity in their language. Hence it is *that they cannot err in the form of their syllogisms*— for we seldom deviate from logical rules, except when we abuse the ambiguity of words, or attribute a different meaning to the middle term in the major and in the minor proposition. It is also the custom of geometers to prefix certain self-evident axioms or principles from which all that they are subsequently to demonstrate flow. Finally, their conclusions are deduced, either from definitions which cannot be called in question, or from those principles and propositions known by the light of nature, which are styled axioms, or from other already established conclusions, which now obtain the cogency of principles. They make not troublesome inquiry into the mood or figure of a syllogism, nor lavish attention on the rules of logic; for such attention, by averting their mind from more necessary objects, would be detrimental rather than advantageous.

Quoted in
Edinburgh Review
Volume 52
January 1836 (p. 228)

Huxley, Aldous
How orderly philosophical is the landscape, are all the inhabitants of this World! It is the creation of a god who 'ever plays the geometer'.

Music at Night and Other Essays
Music At Night

Scaliger, Joseph
A dull and patient intellect such should be your geometers. A great genius cannot be a great mathematician.

Quoted in
Edinburgh Review
Volume 52
January 1836 (p. 229)

Unknown
Is it where the Geometer draws his base,
And eloquent quadrices float through space,
Where the circular points are the open door,
And conics osculate ever more?

The Mathematical Gazette
Volume VIII, Number 117, May 1915 (p. 99)

Smiling flames, it came for him across the chalk lines of the useless hexagram Henry had drawn by mistake instead of the protecting pentagram.
Frederic Brown – (See p. 82)

GEOMETRY

Adler, Irving
Geometry today consists of many subdivisions. There are synthetic geometry, analytic geometry, and differential geometry. There are Euclidean geometry, hyperbolic geometry, and elliptic geometry. There are also metric geometry, affine geometry, projective geometry, and other branches besides. The subdivisions of geometry have been compared to the distinguishable regions within a complex landscape. Most of these regions are in a valley. An explorer who is deep within one region can easily lose sight of the fact that the other regions exist. At a boundary where one region touches another he can see the fact that the regions are related to each other. But seeing the regions pair by pair does not suffice to reveal the pattern of this relationship. There is a path from the valley that leads up the side of a mountain to a clearing at the top. The explorer who reaches this clearing suddenly sees the whole valley laid out before his eyes. From his height at the top of the mountain he can see all the regions of the valley and the pattern that they form . . . he can see the grand design of the valley in all its breathtaking splendor.

A New Look at Geometry (pp. 9–10)

Aristotle
We cannot . . . prove geometrical truths by arithmetic.

Posterior Analytics
Book I, Chapter 7, 35

Bell, Eric T.
The cowboys have a way of trussing up a steer or a pugnacious bronco which fixes the brute so that it can neither move nor think. This is the hog-tie, and it is what Euclid (B.C. 330–275) did to geometry.

The Search for Truth (p. 117)

This would appear to put at least part of the Theory of Demonstration in a category with the efforts of beginners in Geometry: To prove that A equals B: let A equal B; therefore A equals B.

Debunking Science

. . . the only royal road to elementary geometry is ingenuity.

The Development of Mathematics (p. 322)

Berkeley, George
The method of Fluxions [the calculus] is the general key by help whereof the modern mathematicians unlock the secrets of Geometry, and consequently of Nature.

Quoted in E.T. Bell
Men of Mathematics (p. 90)

Bôcher, Maxime
We must, then, admit . . . that there is an independent science of geometry just as there is an independent science of physics, and that either of these *may* be treated by mathematical methods. Thus geometry becomes the simplest of the natural sciences, and its axioms are of the nature of physical laws, to be tested by experience and to be regarded as true only within the limits of the errors of observation.

Bulletin of the American Mathematical Society
2nd Series
The Fundamental Conceptions and Methods in Mathematics
Volume 11, 1904 (p. 124)

Borel, Émile
The goal of geometry is to study those properties of bodies which can be considered independent of their matter, but only with respect to their dimensions and their forms. Geometry measures the surface of a field without bothering to find out whether the soil is good or bad.

Geometry
Premier et Second Cycles (p. 1)

Brown, Frederic
"I've always been poor at geometry," he began . . .

"You're telling *me*," said the demon gleefully.

Smiling flames, it came for him across the chalk lines of the useless hexagram Henry had drawn by mistake instead of the protecting pentagram.

And the Gods Laughed
Naturally

Cedering, Siv
As I picture each planet
floating within the geometric perfections

of space, I think geometry was implanted in man
along with the image of God.
Geometry indeed is God.

Letters from the Floating World
Letters from the Astronomers
II. Johannes Kepler (p. 114)

Chasles, Michel
The doctrines of pure geometry often, and in many questions, give a
simple and natural way to penetrate to the origin of truths, to lay bare
the mysterious chain which unites them, and to make them known
individually, luminously and completely.

Quoted in Morris Kline
Mathematical Thought from Ancient to Modern Times (p. 834)

Comte, Auguste
. . . GEOMETRY is a true natural science;—only more simple, and
therefore more perfect than any other. We must not suppose that, because
it admits the application of mathematical analysis, it is therefore a
purely logical science, independent of observation. Every body studied
by geometers presents some primitive phenomena which, not being
discoverable by reasoning, must be due to observation alone.

The Positive Philosophy
Volume I
Book I, Chapter III (p. 86)

Coolidge, Julian L.
The present author humbly confesses that, to him, geometry is nothing
at all, if not a branch of art . . .

A Treatise on Algebraic Plane Curves
Preface (p. x)

Davies, Paul
Geometry was the midwife of science.

Superforce (p. 165)

de Morgan, Augustus
Geometry, then, is the application of strict logic to those properties of
space and figure which are self-evident, and which therefore cannot be
disputed. But the rigor of this science is carried one step further; for
no property, however evident it may be, is allowed to pass without
demonstration, if that can be given. The question is therefore to
demonstrate all geometrical truths with the smallest possible number
of assumptions.

On the Study and Difficulties of Mathematics (p. 231)

Descartes, René

... we have sufficient evidence that the ancient Geometricians made use of a certain analysis which they extended to the resolution of all problems, though they grudged the secret to posterity.

Rules for the Direction of the Mind
Rule IV

Dieudonné, Jean

... it has been said that the art of geometry is to reason well from false diagrams.

Mathematics—The Music of Reason (p. 37)

Emerson, Ralph Waldo

The astronomer discovers that geometry, a pure abstraction of the human mind, is the measure of planetary motion.

Quoted in Franzo H. Crawford
Introduction to the Science of Physics (p. 33)

Euripides

Mighty is geometry; joined with art, resistless.

Quoted in Stanley Gudder
A Mathematical Journey (p. 67)

Fabre, Jean Henri

Geometry, that is to say, the science of harmony in space, presides over everything. We find it in the arrangement of the scales of a fir-cone, as in the arrangement of an Epeira's lime-snare; we find it in the spiral of a Snail-shell, in the chaplet of a Spider's thread, as in the orbit of a planet; it is everywhere, as perfect in the world of atoms as in the world of immensities.

The Life of the Spider (p. 399)

Galilei, Galileo

SIMPLICO: Indeed, I begin to understand that while logic is an excellent guide in discourse, it does not, as regards stimulation to discovery, compare with the power of sharp distinction which belongs to geometry.

Dialogues Concerning the Two New Sciences
Second Day (p. 190)

Halsted, George Bruce

No mathematical exactness without explicit proof from assumed principles—such is the motto of the modern geometer.

Quoted in Henri Poincaré
The Foundations of Science
The Value of Science
Translator's Introduction (p. 202)

Hamilton, William Rowan
The mathematical process in the symbolical method [i.e., the algebraical] is like running a railroad through a tunneled mountain, that in the ostensive [i.e., the geometrical] like crossing the mountain on foot. The former causes us, by a short and easy transit, to our destined point, but in miasma, darkness and torpidity, whereas the latter allows us to reach it only after time and trouble, but feasting us at each turn with glances of the earth and of the heavens, while we inhale the pleasant breeze, and gather new strength at every effort we put forth.

Quoted in Richard Olson
Scottish Philosophy and British Physics: 1750–1880 (p. 22)

Hermite, Charles
I cannot tell you the efforts to which I was condemned to understand something of the diagrams of Descriptive Geometry, which I detest.

Quoted in E.T. Bell
Men of Mathematics (p. 183)

Herodotus
Sesostris also, . . . made a division of the soil of Egypt among the inhabitants, assigning square plots of ground of equal size to all . . .. If the river carried away any portion of a man's lot, he appeared before the king, and related what had happened; upon which the king sent persons to examine, and determine by measurement the exact extent of the loss . . .. From this practice, I think, geometry first came to be known.

The History of Herodotus
Book II, Section 109

Hobbes, Thomas
Geometry (which is the only Science that it hath pleased God hitherto to bestow on mankind) . . .

Leviathan
Part I, Chapter 4 (p. 26)

. . . Geometry, which is the Mother of all Natural Science . . .

Leviathan
Part IV, Chapter 46 (p. 588)

Huxley, Aldous
. . . a world where beauty and logic, painting and analytic geometry, had become one.

After Many a Summer Dies the Swan
Part I, Chapter III (p. 44)

Kepler, Johannes

Geometry has two great treasures: one is the Theorem of Pythagoras; the other, the division of a line into extreme and mean ratio. The first we may compare to a measure of gold; the second we may name a precious jewel.

Quoted in Carl B. Boyer
A History of Mathematics (p. 55)

Klein, Felix

Projective geometry has opened up for us with the greatest facility new territories in our science, and has rightly been called the royal road to its own particular field of knowledge.

Quoted in E.T. Bell
Men of Mathematics (p. 206)

Kline, Morris

In the house of mathematics there are many mansions and of these the most elegant is projective geometry.

Scientific American
Projective Geometry
Volume 192, Number 1, January 1955 (p. 80)

. . . no branch of mathematics competes with projective geometry in originality of ideas, coordination of intuition in discovery and rigor in proof, purity of thought, logical finish, elegance of proofs and comprehensiveness of concepts. The science born of art proved to be an art.

Scientific American
Projective Geometry
Volume 192, Number 1, January 1955 (p. 86)

Leibniz, Gottfried Wilhelm

Those few things having been considered, the whole matter is reduced to pure geometry, which is the one aim of physics and mechanics.

Quoted in Morris Kline
Mathematical Thought from Ancient to Modern Times (p. 391)

Marvell, Andrew

As lines, so loves oblique may well
Themselves in every angle greet
But ours, so truly parallel,
Though infinite can never meet.

The Poetical Works of Andrew Marvell
Definition of Love
Verse VII

Newton, Sir Isaac
... it is the glory of geometry that from so few principles, fetched from without, it is able to produce so many things.

Mathematical Principles of Natural Philosophy
Preface to the First Edition

Pirsig, Robert M.
One geometry cannot be more true than another; it can only be more *convenient*. Geometry is not true, it is advantageous.

Zen and the Art of Motorcycle Maintenance
Part III, Chapter 22 (p. 264)

Plato
... we are concerned with that part of geometry which relates to war; for in pitching a camp, or taking up a position, or closing or extending the lines of an army, or any other maneuver, whether in actual battle or on a march, it will make all the difference whether a general is or is not a geometrician.

The Republic
Book VII, Section 526

... the knowledge at which geometry aims is knowledge of the eternal, and not of aught perishing and transient.

The Republic
Book VII, Section 527

... geometry will draw the soul towards truth, and create the spirit of philosophy ...

The Republic
Book VII, Section 527

Plotinus
Geometry, the science of the Intellectual entities ...

The Six Enneads
Fifth Ennead IX.11

Poincaré, Henri
Therefore, if there were no solid bodies in nature, there would be no geometry.

The Foundations of Science
Science and Hypothesis
Space and Geometry (p. 73)

... geometry is not true, it is advantageous.

The Foundations of Science
Science and Hypothesis
Experiment and Geometry (p. 91)

That being so what ought one to think of this question: Is the Euclidean Geometry true? The question is nonsense. One might as well ask whether the metric system is true and the old measures false; whether Cartesian co-ordinates are true and polar co-ordinates false.

Nature
Non-Euclidean Geometry
Volume 46, 1891–1892 (p. 407)

Pólya, George
Geometry is the art of correct reasoning on incorrect figures.

How to Solve It (p. 181)

Russell, Bertrand
GEOMETRY, throughout the 17th and 18th centuries, remained, in the war against empiricism, an impregnable fortress of the idealists. Those who held—as was generally held on the Continent—that certain knowledge, independent of experience, was possible about the real world, had only to point to Geometry: none but a madman, they said, would throw doubt on its validity, and none but a fool would deny its objective reference.

An Essay on the Foundations of Geometry
Introduction (p. 1)

... Geometry has been, throughout, of supreme importance in the theory of knowledge.

An Essay on the Foundations of Geometry
Chapter II (p. 54)

All geometrical reasoning is, in the last resort, circular: if we start by assuming points, they can only be defined by the lines or planes which relate them; and if we start by assuming lines or planes, they can only be defined by the points through which they pass.

An Essay on the Foundations of Geometry
Chapter III (p. 120)

It was formerly supposed that Geometry was the study of the nature of space in which we live, and accordingly it was urged, by those who held that what exists can only be known empirically, that Geometry should really be regarded as belonging to applied mathematics. But it has gradually appeared, by the increase of non-Euclidean systems, that Geometry throws no more light upon the nature of space than Arithmetic throws upon the population of the United States.

Mysticism and Logic and Other Essays
Mathematics and the Metaphysicians (p. 92)

Siegel, Eli
A layer cake is geometry and layer cake.

Damned Welcome
Aesthetic Realism Maxims
Part II, #374 (p. 148)

Smith, Henry J.S.
One thing at least they have not forgotten that, geometry is nothing if it be not rigorous, and that the whole educational value of the study is lost, if strictness of demonstration be trifled with. The methods of Euclid are, by almost universal consent, unexceptionable in point of rigour.

Nature
Opening Address by the President Professor Henry J.S. Smith
Section A, Volume 8, September 25, 1873 (p. 450)

Sylvester, James Joseph
He who would know what geometry is, must venture boldly into its depths and learn to think and feel as a geometer. I believe that it is impossible to do this, and to study geometry as it admits of being studied and am conscious it can be taught, without finding the reason invigorated, the invention quickened, the sentiment of the orderly and beautiful awakened and enhanced, and reverence for truth, the foundation of all integrity of character, converted into a fixed principle of the mental and moral constitution, according to the old and expressive adage *"abeunt studia in mores."*

The Collected Mathematical Papers of James Joseph Sylvester
Volume II
A Probationary Lecture on Geometry, Delivered before the
Gresham Committee and the Members of the
Common Council of the City of London, 4 December, 1854 (p. 9)

Geometry may sometimes appear to take the lead over analysis but in fact precedes it only as a servant goes before the master to clear the path and light him on his way.

Philosophic Magazine
Volume 31, 1866 (p. 521)

Thom, René
. . . the spirit of geometry circulates almost everywhere in the immense body of mathematics, and it is a major pedagogical error to seek to eliminate it.

Quoted in A.G. Howson (Editor)
Developments in Mathematical Education: Proceedings of the
Second International Congress on Mathematical Education
Modern Mathematics: Does It Exist (p. 208)

Unknown
Geometry teaches us how to bisex angles.

Source unknown

Veblen, Oswald
At the same time it will not be forgotten that the physical reality of geometry can not be put in evidence with full clarity unless there is an abstract theory also . . . Thus, for example, while the term electron may have more than one physical meaning, it is by no means such a protean object as a point or a triangle.

Science
Geometry and Physics
Volume LVII, Number 1466, February 2, 1923 (p. 134)

Voltaire
. . . but of all the sciences, the most absurd, and that which in my opinion, is most calculated to stifle genius of every kind, is geometry. The objects about which this ridiculous science is conversant, are surfaces, lines, and points, that have no existence in nature.

The Best Known Works of Voltaire
Jeannot and Colin (p. 287)

. . . the geometrician makes a hundred thousand curved lines pass between a circle and a right line that touches it, when, in reality, there is not room for a straw to pass there.

The Best Known Works of Voltaire
Jeannot and Colin (p. 287)

Geometry, if we consider it in its true light, is a mere jest, and nothing more.

The Best Known Works of Voltaire
Jeannot and Colin (p. 287)

Did anyone ever so much as think of talking geometry in good company?

The Best Known Works of Voltaire
Jeannot and Colin (p. 288)

Warner, Sylvia Townsend
Geometry would be best to begin with, plain plane geometry, immutably plane. Surely if anything could minister to the mind diseased it would be the steadfast contemplation of a right angle, an existence that no mist of human tears could blur, no blow of fate deflect.

Mr. Fortune's Maggot (p. 162)

Weyl, Hermann
Geometry became one of the most powerful expressions of that sovereignty of the intellect that inspired the thought of those times. At a later epoch, when the intellectual despotism of the church, which had been maintained through the middle ages, had crumbled, and a wave of skepticism threatened to sweep away all that had seemed most fixed,

those who believed in truth clung to Geometry as to a rock, and it was the highest ideal of every scientist to carry on his science *"more geometrico."*

Space, Time, Matter
Introduction (p. 1)

Wheeler, John A.
There is nothing in the world except curved empty space. Geometry bent one way here describes gravitation. Rippled another way somewhere else it manifests all the qualities of an electromagnetic wave. Excited at still another place, the magic material that is space shows itself as a particle. There is nothing that is foreign and "physical" immersed in space. Everything that is, is constructed out of geometry.

Quoted by Cecil M. DeWitt and John A. Wheeler in
Battelle Rencontres (p. 273)
1967 Lectures in Mathematics and Physics

Whewell, William
This science is one of indispensable use and constant reference, for every student of the laws of nature; for the relations of space and number are the alphabet in which those laws are written. But besides the interest and importance of this kind which geometry possesses, it has a great and peculiar value for all who wish to understand the foundations of human knowledge, and the methods by which it is acquired. For the student of geometry acquires, with a degree of insight and clearness which the unmathematical reader can but feebly imagine, a conviction that there are necessary truths, many of them of a very complex and striking character; and that a few of the most simple and self-evident truths which it is possible for the mind of man to apprehend, may, by systematic deduction, lead to the most remote and unexpected results.

The Philosophy of the Inductive Sciences
Volume I
Part 1, Book 2, Chapter 4, section 8 (pp. 99–100)

Wordsworth, William
But who shall parcel out
His intellect by geometric rules,
Split like a province into round and square?

The Prelude
School-Time (Continued) (p. 41)

HYPERBOLA

Carroll, Lewis

What mathematician has ever pondered over an hyperbola, mangling the unfortunate curve with lines of intersection here and there, in his efforts to prove some property that perhaps after all is a mere calumny, who has not fancied at last that the ill-used locus was spreading out its asymptotes as a silent rebuke, or winking one focus at him in contemptuous pity?

The Complete Works of Lewis Carroll
The Dynamics of a Parti-cle (p. 1130)

Frere, C.
Canning, B.

Not thus HYPERBOLA;—with subtlest art
The blue-eyed wanton plays her changeful part;
Quick as her *conjugated axes* move
Through every posture of luxurious love,
Her supportive limbs with easiest grace expand; . . .

Quoted in Charles Edmonds
Poetry of the Anti-Jacobin
The Loves of the Triangle
Canto II, l. 115–19

HYPOTHESIS

Asquith, Herbert
Jolie hypothèse quelle explique tant de choses.
[A pretty hypothesis which explains so many things.]

Speech in the House of Commons
March 29, 1917

Baez, Joan
. . . hypothetical questions get hypothetical answers.

Daybreak
What Would You Do If (p. 134)

Barry, Frederick
Hypothesis, however, is an inference based on knowledge which is insufficient to prove its high probability.

The Scientific Habit of Thought
The Elements of Theory (p. 164)

Bruner, Jerome Seymore
The shrewd guess, the fertile hypothesis, the courageous leap to a tentative conclusion—these are the most valuable coin of the thinker at work.

The Process of Education (p. 14)

Carroll, Lewis
"Would you tell me, please, which way I ought to go from here?"

"That depends a good deal on where you want to get to," said the Cat.

"I don't much care where——" said Alice.

"Then it doesn't matter which way you go," said the Cat.

The Complete Works of Lewis Carroll
Alice's Adventures in Wonderland
Pig and Pepper

Cohen, Morris
There is . . . no genuine progress in scientific insight through the Baconian method of accumulating empirical facts without hypotheses or anticipation of nature. Without some guiding idea we do not know what facts to gather . . . we cannot determine what is relevant and what is irrelevant.

A Preface to Logic (p. 148)

Cort, David
But suspicion is a thing very few people can entertain without letting the hypothesis turn, in their minds, into fact . . .

Social Astonishments
ONE
Believing in Books

Evans, Bergen
An honorable man will not be bullied by a hypothesis.

The Natural History of Nonsense
A Tale of a Tub

We see what we want to see, and observation conforms to hypothesis.

The Natural History of Nonsense
A Tale of a Tub

Fabing, Harold
Mar, Ray
Many confuse hypothesis and theory. An hypothesis is a possible explanation; a theory, the correct one.

Fischerisms (p. 7)

Freud, Sigmund
In the complete absence of any theory of the instincts which would help us to find our bearings, we may be permitted, or rather, it is incumbent upon us, in the first place to work out any hypothesis to its logical conclusion, until it wither fails or becomes confirmed.

On Narcissism
I (p. 401)

Goethe, Johann Wolfgang von
Hypotheses are the scaffolds which are erected in front of a building and removed when the building is completed. They are indispensable to the worker; but he must not mistake the scaffolding for the building.

Goethe's Poems and Aphorisms (p. 197)

Hypotheses are lullabies with which the teacher lulls his pupils to sleep. The thinking and faithful observer learns to know his limitation more and more; he sees that the further knowledge extends the more problems arise.

Goethe's Poems and Aphorisms (p. 197)

Holmes, Sherlock

"If the fresh facts which come to our knowledge all fit themselves into the scheme, then our hypothesis may gradually become a solution."

In Arthur Conan Doyle's
The Complete Sherlock Holmes
The Adventure of Wisteria Lodge

Huxley, Thomas H.

. . . the great tragedy of Science—the slaying of a beautiful hypothesis by an ugly fact . . .

Collected Essays
Volume VIII
Biogenesis and Abiogenesis (p. 244)

. . . it is the first duty of a hypothesis to be intelligible . . .

Man's Place in Nature
II (p. 126)

Machover, Maurice

My name is $\aleph$ (known as c),
You have no upper bound for me.
I go as low an $\aleph_1$,
And then soar back restrained by none.
The hierarchies of steps I roam,
And almost every step, my home.
Of course J. König has ordained
That certain first steps can't be gained.
But Cohen and Gödel set me free,
$\aleph_\Omega$ I can be.
You cannot catch me in your net,
Discreteness hasn't trapped me yet.
So learn the moral of my tale,
I cannot fit into a scale.
For any scale you think will serve,
Might press me down, but then I'll swerve.

Mathematics Magazine
Ode to the Continuum Hypothesis
Volume 50, Number 2, March 1977 (p. 94)

Sterne, Laurence

It is the nature of an hypothesis, when once a man has conceived it, that it assimilates every thing to itself, as proper nourishment; and, from the first moment of your begetting it, it generally grows stronger by every thing you see, hear, read, or understand. This is of great use.

Tristram Shandy
Book 2, Chapter 19

Unknown

[Hypothesis] Something usually murdered by facts.

Source unknown

Infinity is a floorless room without walls or ceiling.
Unknown - (See p. 108)

INDUCTION

Laplace, Pierre Simon
Analysis and natural philosophy owe their most important discoveries to this fruitful means, which is called *induction*. Newton was indebted to it for his theorem of the binomial and the principle of universal gravity.

A Philosophical Essay on Probability (p. 176)

Even in the mathematical sciences, our principal instruments to discover the truth are induction and analogy.

Oeuvres Complètes de Laplace
Introduction
Volume 7 (p. v)

Lewis, C.S.
This is called the inductive method. Hypothesis, my dear young friend, establishes itself by a cumulative process: or, to use popular language, if you make the same guess often enough it ceases to be a guess and becomes a Scientific Fact.

*The Pilgrim's Regress: An Allegorical Apology for Christianity,
Reason and Romanticism* (p. 37)

Mill, John Stuart
There is in every step of an arithmetical or algebraical calculation a real induction, a real inference from facts to facts, and what disguises the induction is simply its comprehensive nature, and the consequent extreme generality of its language.

System of Logic
Book 2, Chapter 2, 2

Unknown
Inductive Process:
Formulate hypothesis
Apply for grant

Perform experiments or gather data to test hypothesis
Alter data to fit hypothesis
Publish

Source unknown

Whitehead, Alfred North

. . . the chief reason in favour of any theory on the principles of mathematics must always be inductive, *i.e.*, it must lie in the fact that the theory in question enables us to deduce ordinary mathematics. In mathematics, the greatest degree of self-evidence is usually not to be found quite at the beginning, but at some later point; hence the early deductions, until they reach this point, give reasons rather from them, than for believing the premises because true consequences follow from them, than for believing the consequences because they follow from the premises.

Principia Mathematica
Preface (p. v)

INFINITE

Anaxagoras
There is no smallest among the small and no largest among the large,
But always something still smaller and something still larger.

Quoted in Eli Maor
To Infinity and Beyond: A Cultural History of the Infinite (p. 2)

Aristotle
But nature flies from the infinite, for the infinite is unending or imperfect,
and Nature ever seeks an end.

Generation of Animals
Book I, 715b15

Our account does not rob the mathematicians of their science, by
disproving the actual existence of the infinite in the direction of increase,
in the sense of the untracersable. In point of fact they do not need the
infinite and do not use it. They postulate only that the finite straight line
may be produced as far as they wish.

Physics
Book III, Chapter 7, 27

Aurelius, Marcus [Antoninus]
Infinity is a fathomless gulf, into which all things vanish.

Quoted in Eli Maor
To Infinity and Beyond: A Cultural History of the Infinite (p. vii)

Bartlett, Elizabeth
Because I longed
to comprehend the infinite
I drew a line
between the known and unknown.

In Ernest Robson and Jet Wimp
Against Infinity
Because I Longed

Beerbohm, Max
The attempt to conceive Infinity had always been quite arduous enough for me.

Mainly on the Air
A Note on the Einstein Theory (p. 137)

Bell, Eric T.
The toughminded suggest that the theory of the infinite elaborated by the great mathematicians of the Nineteenth and Twentieth Centuries, without which mathematical analysis as it is actually used today is impossible, has been committing suicide in an unnecessarily prolonged and complicated manner for the past half century.

Debunking Science

Berkeley, George
Of late the speculations about Infinities have run so high, and grown to such strange notions, as have occasioned no small scruples and disputes among the geometers of the present age. Some there are of great note who, not content with holding that finite lines may be divided into an infinite number of parts, do yet farther maintain that each of those infinitesimals is itself subdivisible into an infinity of other parts or infinitesimals of a second order, and so on ad infinitum. These, I say, assert there are infinitesimals of infinitesimals of infinitesimals, &c., without ever coming to an end; so that according to them an inch does not barely contain an infinite number of parts, but an infinity of an infinity of an infinity ad infinitum of parts.

The Principles of Human Knowledge
Section 130

Bing, Ilse
Infinitesimal is the nearest to zero
infinitesimal is so small
that it is no longer something
but it is not yet nothing

Quoted in Ernest Robson and Jet Wimp
Against Infinity
Infinitesimal

Blake, William
Too see the World in a grain of sand,
And a Heaven in a wild flower:
Hold Infinity in the palm of your hand,
And Eternity in an hour.

BLAKE: The Complete Poems
The Pickering Manuscript, Auguries of Innocence, l. 1–4

Borges, Jorge Luis
The ignorant suppose that infinite number of drawings require an infinite amount of time; in reality it is quite enough that time be infinitely subdivisible, as is the case in the famous parable of the Tortoise and the Hare. This infinitude harmonizes in an admirable manner with the sinuous numbers of Chance and of the Celestial Archetype of the Lottery, adored by the Platonists.

Ficciones
The Babylon Lottery (p. 70)

Box, G.E.P.
It is a pity, therefore, that the authors have confined their attention to the relatively simple problem of determining the approximate distribution of arbitrary criteria and have failed to produce any sort of justification for the tests they propose. In addition to those functions studied there are an infinity of others, and unless some principle of selection is introduced we have nothing to look forward to but an infinity of test criteria and an infinity of papers in which they are described.

Journal of the Royal Statistical Society
Discussion
Ser. B, Volume 18, Number 1, 1956 (p. 29)

Buzz Lightyear
To infinity and beyond!

From the movie *Toy Story*

Camus, Albert
Somebody has to have the last word. Otherwise, every reason can be met with another one and there would never be an end to it.

The Fall (p. 45)

Carlyle, Thomas
The moment of discovery, 'spontaneous illumination . . .' The infinite is made to blend itself with the finite, to stand visible, as it were, attainable there.

Quoted in Roger A. MacGowan and Frederick I. Ordway, III
Intelligence in the Universe (p. 49)

Carus, Paul
Infinity is the land of mathematical hocus pocus. There Zero the magician is king. When Zero divides any number he changes it without regard to its magnitude into the infinitely small; and inversely, when divided by any number he begets the infinitely great.

The Monist
Logical and Mathematical Thought
Volume 20, Number 1, January 1910 (p. 69)

Cousins, Norman
Infinity converts the possible into the inevitable.

Saturday Review
Editor's Odyssey
April 15, 1978 (p. 18)

Crane, Hart
But the star-glistered salver of infinity,
The circle, blind crucible of endless space,
Is sluiced by motion,—subjugated never.

The Collected Poems of Hart Crane
The Bridge
Cape Hatteras

da Vinci, Leonardo
What is that thing which does not give itself, and which if it were to give
itself would not exist? It is the infinite . . .

The Notebooks of Leonardo da Vinci
Volume I
Mathematics (p. 627)

de Morgan, Augustus
Great fleas have little fleas upon their backs to bite 'em,
And little fleas have lesser fleas, and so ad infinitum.
And the great fleas themselves, in turn have greater fleas to go on;
While these again have greater still, and greater still, and so on.

Budget of Paradoxes
Volume II (p. 191)

I had expressed my wish to have a *thermometer of probability*, with
impossibility at one end, as 2 and 2 make 5, and necessity at the other
as 2 and 2 make 4 . . .

Budget of Paradoxes
Volume II (p. 246)

de Musset, Alfred
I cannot help it,—in spite of myself, infinity torments me.

L'Espoir en Dieu

Dryden, John
But how can finite grasp Infinity?

The Poetical Works of Dryden
Hind and the Panther
I, l. 105

Eldridge, Paul
Truth must be judged in terms of time and space; superstition in terms of eternity and infinity.

Maxims for a Modern Man
2533

Emerson, Ralph Waldo
. . . and thus ever, behind the coarse effect, is a fine cause, which, being narrowly seen, is itself the effect of a finer cause.

The Selected Writings of Ralph Waldo Emerson
Essays
First Series
Circles (p. 280)

Froude, James Anthony
Large forms resolve themselves into parts, down so far as we can see into infinity.

Short Studies on Great Subjects
Calvinism (p. 16)

Gamow, George
One, two, three—infinity

Title of Book

Gauss, Carl Friedrich
Infinity is only a figure of speech, meaning a limit to which certain ratios may approach as closely as desired, when others are permitted to increase indefinitely.

Quoted in Eli Maor
To Infinity and Beyond: A Cultural History of the Infinite (p. ix)

Gleick, James
In the mind's eye, a fractal is a way of seeing infinity.

Chaos: Making a New Science (p. 98)

Harrison, Edward
Only a cosmic jester could perpetrate eternity and infinity . . .

Masks of the Universe (p. 201)

If eternity is stillness, then infinity of space is sheer madness.

Masks of the Universe (p. 202)

Hawking, Stephen
In an infinite universe, every point can be regarded as the center, because every point has an infinite number of stars on each side of it.

A Brief History of Time (p. 5)

Hilbert, David
The infinite, like no other problem, has always deeply moved the soul of men. The infinite, like no other idea, has had a stimulating and fertile influence upon the mind. But the infinite is also more than any other concept, in need of clarification.

Quoted in Hermann Weyl
Philosophy of Mathematics and Natural Science (p. 66)

Hobbes, Thomas
When we say anything is infinite, we signify only that we are not able to conceive the ends and bounds of the thing named.

Quoted in Eli Maor
To Infinity and Beyond: A Cultural History of the Infinite (p. viii)

Kant, Immanuel
But the infinite is absolutely (not merely comparatively) great. In comparison with this all else (in the way of magnitudes of the same order) is small. But the point of capital importance is that the mere ability even to think it as a whole indicates a faculty of mind transcending every standard sense. For the latter would entail a comprehension yielding as unit a standard bearing to the infinite ration expressible in numbers, which is impossible.

The Critique of Judgment
Critique of Aesthetic Judgment
Section 26

Kasner, Edward
Newman, James R.
With the Hottentots, infinity begins at three.

Mathematics and the Imagination (p. 19)

Locke, John
He that thinks he has a positive idea of infinite space, will, when he considers it, find that he can no more have a positive idea of the greatest, than he has of the least space. For in this latter, which seems the easier of the two, and more within our comprehension, we are capable only of a comparative idea of smallness, which will always be less than any one whereof we have the positive idea. All our positive ideas of any quantity, whether great or little, have always bounds, though our comparative idea, whereby we can always add to the one, and take from the other, hath no bounds. For that which remains, either great or little, not being comprehended in that positive idea which we have, lies in obscurity; and we have no other idea of it, but of the power of enlarging the one and diminishing the other, without ceasing. A pestle and mortar will as soon bring any particle of matter to indivisibility, as the acutest thought of a

mathematician; and a surveyor may as soon with his chain measure out infinite space, as a philosopher by the quickest flight of mind reach it, or by thinking comprehend it; which is to have a positive idea of it. He that thinks on a cube of an inch diameter, has a clear and positive idea of it in his mind, and so can frame one of 1/2, 1/4, 1/8, and so on, till he has the idea in his thoughts of something very little; but yet reaches not the idea of that incomprehensible littleness which division can produce. What remains of smallness is as far from his thoughts as when he first began; and therefore he never comes at all to have a clear and positive idea of that smallness which is consequent to infinite divisibility.

An Essay Concerning Human Understanding
Book II, Chapter 17, Section 18

Loomis, Elisha S.
In the use of this method (of infinities) the pupil must be awake and thinking, for when the infinite is employed in an argument by the unskilled, the conclusion is often most absurd.

Quoted in Eli Maor
To Infinity and Beyond: A Cultural History of the Infinite (p. 34)

Lucretius
Again if for the moment all existing space be held to be bounded, supposing a man runs forward to its outside borders, and stands on the utmost verge and then throws a winged javelin, do you choose that when hurled with vigorous force it shall advance to the point to which it has been sent and fly to a distance, or do you decide that something can get in its way and stop it? for you must admit and adopt one of the two suppositions; either of which shuts you out from all escape and compels you to grant that the universe stretches without end.

On the Nature of Things
Book I, l. 968

Milton, John
Infinity is a dark illimitable ocean without bound.

Quoted in Eli Maor
To Infinity and Beyond: A Cultural History of the Infinite (p. viii)

Pascal, Blaise
We know that there is an infinite, and we are ignorant of its nature.

Pensées
Section III, 233

Unity joined to infinity adds nothing to it, no more than one foot to an infinite measure. The finite is annihilated in the presence of the infinite, and becomes a pure nothing.

Pensées
Section III, 233

Phrase, Latin
Ad infinitum
[To infinity]

Pierpont, James
. . . the notion of infinity is our greatest friend; it is also the greatest enemy of our peace of mind . . . Weirstrass taught us to believe that we had at last thoroughly tamed and domesticated this unruly element. Such however is not the case; it has broken loose again. Hilbert and Brouwer have set out to tame it once more. For how long? We wonder.

Bulletin of the American Mathematical Society
Mathematical Rigor
Volume 34, January–February 1928 (p. 47)

Richardson, Lewis
Big whorls have little whorls
Which feed on their velocity,
And little whorls have lesser whorls,
And so on to viscosity.

Quoted in Ian Stewart
Does God Play Dice? (p. 196)

Royce, Josiah
. . . let us suppose, if you please, that a portion of the surface of England is very perfectly levelled and smoothed, and is then devoted to the production of our precise map of England. That in general then, should be found upon the surface of England, map constructions which more or less roughly represent the whole of England,—all this has nothing puzzling about it . . . But now suppose that this our resemblance is to be made absolutely exact . . . A map of England, contained within England, is to represent, down to the minutest detail, every contour and marking, natural or artificial, that occurs upon the surface of England . . .

One who, with absolute exactness of perception, looked down upon the ideal map thus supposed to be constructed, would see lying upon the surface of England, and at a definite place thereon, a representation of England on as large or small a scale as you please . . . This representation, which would repeat in the outer portions the details of the former, but upon a smaller space, would be seen to contain yet another England, and this another, and so on without limit.

The World and the Individual
Supplementary Essay
Section III, Part I (pp. 504–5)

Rucker, Rudy
The study of infinity is much more than a dry academic game. The intellectual pursuit of the Absolute Infinite is, as Georg Cantor realized, a form of the soul's quest for God. Whether or not the goal is ever reached, an awareness of the process brings enlightenment.

Infinity and the Mind
Preface (p. ix)

Shakespeare, William
. . . I could be bounded in a nutshell and count myself a king of infinite space . . .

Hamlet, Prince of Denmark
Act II, Scene 2, l. 263

Shelley, Percy Bysshe
. . . infinity within,
Infinity without . . .

The Complete Poetical Works of Shelley
Queen Mab, l. 22

Swift, Jonathan
So, Nat'ralists observe, a Flea
Hath smaller Fleas that on him prey,
And these have smaller Fleas to bite 'em
And so proceed, ad infinitum.

The Portable Swift
On Poetry—A Rhapsody

Tolstoy, Leo
And so to imagine the action of a man entirely subject to the law of inevitability without any freedom, we must assume the knowledge of an infinite number of space relations, an infinitely long period of time, and an infinite series of causes.

War and Peace
Second Epilogue, Chapter X

Arriving at infinitesimals, mathematics, the most exact of sciences, abandons the process of analysis and enters on the new process of the integration of unknown, infinitely small, quantities.

War and Peace
Second Epilogue, Chapter XI

Unknown
Possibilities are infinite.

Source unknown

There was a young man from Trinity,
Who solved the square root of infinity.
　While counting the digits,
　He was seized by the fidgets,
Dropped science, and took up divinity.

Reproduced in Helen Plotz
Imagination's Other Place
There Was A Young Man From Trinity (p. 80)

Infinity is where things happen that don't.

Quoted in W.W. Sawyer
Prelude to Mathematics (p. 143)

Infinity is that dimension without end which the human mind cannot grasp.

Quoted in Eli Maor
To Infinity and Beyond: A Cultural History of the Infinite (p. xiii)

Infinity is a floorless room without walls or ceiling.

Quoted in Eli Maor
To Infinity and Beyond: A Cultural History of the Infinite (p. xiii)

Wittgenstein, Ludwig
"Ought the word 'infinite' to be avoided in mathematics?" Yes; where it appears to confer a meaning upon the calculus; instead of getting one from it.

Remarks on the Foundations of Mathematics
Appendix II, 17 (p. 63e)

INTEGERS

Graham, Ronald L.
The trouble with integers is that we have examined only the very small ones. Maybe all the exciting stuff happens at really big numbers, ones we can't even begin to think about in any very definite way. Our brains have evolved to get us out of the rain, find where the berries are, and keep us from getting killed. Our brains did not evolve to help us grasp really large numbers or to look at things in a hundred thousand dimensions.

Quoted in Clifford A. Pickover
Keys to Infinity (p.xiv)

Minkowski, Hermann
Integers are the fountainhead of all mathematics.

Diophantische Approximationen: eine Einführung in die Zahlentheorie,
von Hermann Minkowski
Preface

INTEGRATION

de Morgan, Augustus
Common integration is only the *memory of differentiation*, the different artifices by which integration is effected are changes, not from the known to the unknown, but from forms in which memory will not serve us to those in which it will.

Transactions of the Cambridge Philosophical Society
Volume 8, 1844 (p. 188)

Kettering, Charles Franklin
. . . the most highly satisfactory use of the reverse-curve sign of integration used in calculus is for those two S-openings in the top of a violin.

Quoted in T.A. Boyd
Professional Amateur (p. 209)

Rankine, W.J.M.
Now integrate L with respect to dt,
   "(t Standing for time and persuasion);
"Then, between proper limits, 'tis easy to see,
"The definite integral Marriage must be—
   "(A very concise demonstration)."

Songs and Fables
The Mathematician in Love
Verse 7

Schenck, Hilbert Jr.
"Oh, hast thou solved the integral?
Here is a raise, my brainish boy!"
He threw his time cards in the air
And clapped his hands with joy.

The Magazine of Fantasy and Science Fiction
Wockyjabber
May 1960

Unknown
Integral z-squared dz
From 1 to the square root of 3
Times the cosine
Of three pi over 9
Equals log of the cube root of 'e'

Source unknown

Mathematics is armchair science.
Chandler Davis – (See p. 193)

IRRATIONAL NUMBERS

Bell, Eric T.
Roughly it amounts to this: mathematical analysis as it works today must make use of irrational numbers (such as the square root of two); the sense if any in which such numbers exist is hazy. Their reputed mathematical existence implies the disputed theories of the infinite. The paradoxes remain. Without a satisfactory theory of irrational numbers, among other things, Achilles does not catch up with the tortoise, and the earth cannot turn on its axis. But as Galileo remarked, it does. It would seem to follow that something is wrong with our attempts to compass the infinite.

Debunking Science

Dantzig, Tobias
To attempt to apply rational arithmetic to a problem in geometry resulted in the first crisis in the history of mathematics. The two relatively simple problems—the determination of the diagonal of a square and that of the circumference of a circle—revealed the existence of new mathematical beings for which no place could be found within the rational domain.

Quoted in Eli Maor
To Infinity and Beyond: A Cultural History of the Infinite (p. 44)

Leibniz Gottfried Wilhelm
. . . a miracle of analysis, a monster of the ideal world, almost an amphibian between being and not being.

Quoted in Walter R. Fuchs
Mathematics for the Modern Mind (p. 168)

Zamyatin, Yevgeny
One day Plapa told us about irrational numbers, and, I remember, I cried, banged my fists on the table, and screamed, "I don't want $\sqrt{-1}$! Take $\sqrt{-1}$ out of me!" This irrational number had grown into me like something foreign, alien, terrifying. It devoured me—it was impossible to conceive, to render harmless, because it was outside *ration*.

We
Eighth Entry (p. 36)

KNOTS

Carroll, Lewis
"A knot," said Alice. "Oh, do let me help to undo it!"

The Complete Works of Lewis Carroll
A Tangled Tale
Appendix (p. 1025)

LIMIT

Lehrer, Tom
There's a delta for every epsilon,
It's a fact that you can always count upon.
There's a delta for every epsilon
   And now and again,
   There's also an N.

But one condition I must give:
The epsilon must be positive
A lonely life all the others live,
   In no theorem
   A delta for them.

How sad, how cruel, how tragic,
How pitiful, and other adjec-
Tives that I might mention.
The matter merits our attention.

If an epsilon is a hero,
Just because it is greater than zero,
It must be mighty discouragin'
To lie to the left of the origin.

This rank discrimination is not for us,
We must fight for an enlightened calculus,
Where epsilons all, both minus and plus,
   Have deltas
   To call their own.

American Mathematical Monthly
There's A Delta For Every Epsilon
Volume 81, Number 6, June–July 1974 (p. 612)

Merriman, Gaylord M.
The limit concept is not armchair fantasy, dissolving with the pipesmoke of its dreamer. It is the stuff of life.

To Discover Mathematics
Chapter 9 (p. 254)

Unknown

$$\lim_{n \to \infty} \frac{\sin(x)}{n} = 6$$

Proof: cancel the *n* in the numerator and the denominator.

Source unknown

I THINK THAT'S QUITE ENOUGH
LOGIC FOR ONE EVENING..!

But, logic, like whiskey, loses its beneficial effect when taken in too large quantities.

Lord Dunsany – (See p. 119)

LINE

Bell, Eric T.
The straight line of the geometers does not exist in the material universe. It is a pure abstraction, an invention of the imagination or, if one prefers, an idea of the Eternal Mind.

The Magic of Numbers (p. 57)

Emerson, Ralph Waldo
Gave his sentiment divine
Against the being of a line.
Line in Nature is not found;
Unit and universe are round;
In vain produced, all rays return . . .

The Complete Works of Ralph Waldo Emerson
Volume IX
Uriel

Newton, Sir Isaac
We ought either to exclude all lines, beside the circle and right line, out of geometry, or admit them according to the simplicity of their descriptions, in which case the Conchoid yields to none except the circle.

That is *arithmetically* more simple, which is determined by the more simple equations, but that is *geometrically* more simple, which is determined by the more simple drawing of lines . . .

Quoted in William Allen
American Journal of Science
On the Curves of Trisection
Volume 4, 1822 (p. 344)

LOGARITHM

Graham, L.A.
Mary had a little lamb
Whose fleece in spirals grew;
She, being quite perceptive said,
"They're logarithmic, too.
Since a's one inch, the length of wool
At any time you see,
Is merely 1.414
Times the radian power of e."

Ingenious Mathematical Problems and Methods
Mathematical Nursery Rhyme No. 16

LOGIC

Boutroux, Pierre
Logic is invincible because in order to combat logic it is necessary to use logic.

Quoted in Morris Kline
Mathematical Thought from Ancient to Modern Times (p. 1182)

Carroll, Lewis
"I know what you're thinking about," said Tweedledum; "but it isn't so, nohow."

"Contrariwise," continued Tweedledee, "if it was so, it might be; and if it were so, it would be; but as it isn't, it ain't. That's logic."

The Complete Works of Lewis Carroll
Through the Looking Glass
Tweedledum and Tweedledee

Chesterton, Gilbert Keith
A great deal is said in these days about the value or valuelessness of logic. In the main, indeed, logic is not a productive tool so much as a weapon for defense. A man building up an intellectual system has to build like Nehemiah, with the sword in one hand and the trowel in the other. The imagination, the constructive quality, is the trowel, and the argument is the sword. A wide experience of actual intellectual affairs will lead most people to the conclusion that logic is mainly valuable as a weapon wherewith to exterminate logicians.

The G.K. Chesterton Calendar
January Ten

Clough, Arthur Hugh
Good, too, Logic, of course; in itself, but not in fine weather.

Quoted in James R. Newman
The World of Mathematics
Volume III
The Bothie of Tober-na-vuolich (p. 1878)

Colton, Charles Caleb
Logic is a large drawer, containing some useful instruments, and many more that are superfluous. A wise man will look into it for two purposes, to avail himself of those instruments that are really useful, and to admire the ingenuity with which those that are not so, are assorted and arranged.

Lacon (p. 163)

Dunsany, Lord Edward John Moreton Drax Plunkett
But, logic, like whiskey, loses its beneficial effect when taken in too large quantities.

My Ireland
Weeds & Moss (p. 186)

Heaviside, Oliver
Logic can be patient for it is eternal.

Quoted in Morris Kline
Mathematical Thought from Ancient to Modern Times (p. 3)

Holmes, Oliver Wendell
Logic is logic. That's all I say.

The Autocrat of the Breakfast-Table
Chapter 11

Jerome, Jerome K.
When a twelfth century youth fell in love, he did not take three paces backward, gaze into her eyes, and tell her she was too beautiful to live. He said he would step outside and see about it. And if, when he got out, he met a man and broke his head—the other man's head, I mean—then that proved that his—the first fellow's girl—was a pretty girl. But if the other fellow broke his head—not his own you know, but the other fellow's—the other fellow to the second fellow, that is, because of course the other fellow would only be the other fellow to him, not the first fellow who—well, if he broke his head, then his girl— not the other fellow's, but the fellow who was the—Look here, if A broke B's head, then A's girl was a pretty girl: but if B broke A's head, then A's girl wasn't a pretty girl, but B's girl was.

The Idle Thoughts of an Idle Fellow
On Being Idle

Jones, Raymond F.
Logic hasn't wholly dispelled the society of witches and prophets and sorcerers and soothsayers.

The Non-Statistical Man (p. 85)

Jowett, Benjamin

Logic is neither a science nor an art, but a dodge.

Quoted in James R. Newman
The World of Mathematics
Volume IV (p. 2402)

Russell, Bertrand

. . . logic is the youth of mathematics . . .

Introduction to Mathematical Philosophy
Mathematics and Logic (p. 194)

Unknown

Reiteration of an argument is often more effective than its inherent logic.

Source unknown

Whitehead, Alfred North

Logic, properly used, does not shackle thought. It gives freedom, and above all, boldness. Illogical thought hesitates to draw conclusions, because it never knows either what it means, or what it assumes, or how far it trusts its own assumptions, or what will be the effect of any modification of assumptions.

The Organization of Thought (p. 132)

Neither logic without observation, nor observation without logic, can move one step in the formation of science.

The Organization of Thought (p. 132)

Wittgenstein, Ludwig

In der Logik ist nichts zufällig.
[Nothing, in logic, is accidental.]

Tractatus Logico-Philosophicus
2.012 (p. 31)

The disastrous invasion of mathematics by logic.

Remarks on the Foundations of Mathematics
Appendix IV, 24 (p. 145e)

LOGICIAN

Futrelle, Jacques

First and above all he was a logician. At least thirty-five years of the half-century or so of his existence had been devoted exclusively to proving that two and two always equal four, except in unusual cases, where they equal three or five, as the case may be.

Best "Thinking Machine" Detective Stories
The Problem of Cell 13

MAP

Bell, Eric T.
The map is not the thing mapped. When the map is identified with the thing mapped we have one of the vast melting pots of numerology.

Numerology

Boehm, G.A.W.
The question of what properties, such as angle or area, are reproduced on a map without distortion is of prime interest to mathematicians. The question extends far beyond the confines of geometry, for all mathematics can be considered broadly as a study of maps and mapping.

The New World of Mathematics (p. 124)

Carroll, Lewis
He had bought a large map representing the sea
  Without the least vestige of land:
And the crew were much pleased when they found it to be
  A map they could all understand.

The Complete Works of Lewis Carroll
The Hunting of the Snark
Fit the Second
The Bellman's Speech

Francis, Richard L.
Guthrie shaded the map's every section,
of Albion's fair isle with affection.
Yet his pencils grew duller,
Drawing regions to color,
So de Morgan he sought for direction.
A "quaternion" of color embraces,
Maps Hamilton fiercely retraces,
Oh conjecture so nice,

MAP 123

Four colors suffice,
Whether England or faraway places.

Mathematics Magazine
On Coloring a Map
Volume 65, Number 5, December 1990 (p. 327)

Sholander, Marlow
Within your lifetime will, perhaps,
As souvenirs from distant suns
Be carried back to earth some maps
Of planets and you'll find that one's
So hard to color that you've got
To use five crayons. Maybe, not.

Mathematics Magazine
Maybe (p. 20)
Volume 35, Number 1, January 1962

SEE LADS –

WE ARE NOT LOST

HERE BE THE MAP...

He had bought a large map representing the sea
Without the least vestige of land:
And the crew were much pleased when they found it to be
A map they could all understand.
Lewis Carroll – (See p. 122)

MATHEMATICAL

Aristotle

The chief forms of beauty are order and symmetry and definiteness, which the mathematical sciences demonstrate in a special degree.

Metaphysica
Book VIII, Chapter 3, l. 1078b

Barry, Frederick

No amount of logical or mathematical investigation alone, moreover, can ever establish a fact.

The Scientific Habit of Thought (p. 111)

Bell, Eric T.

No satisfactory justification has ever been given for connecting in any way the consequences of mathematical reasoning with the physical world.

Debunking Science

When a complicated mathematical argument ends in a spectacular prediction, subsequently verified by observation or experiment, a physicist may be excused for feeling that he has participated in a miracle. And when a skilled mathematician astounds himself with a discovery he had no conscious intention of striving after, he may well believe for a few moments as Pythagoras believed all his life, and may even repeat— after the eminent English mathematician, G.H. Hardy—the following confession of faith. "I believe that mathematical reality lies outside us, that our function is to discover or *observe* it, and that the theorems which we prove, and which we describe grandiloquently as our 'creations,' are simply our notes of our observations. This view has been held, in one form or another, by many philosophers of high reputation from Plato onwards . . ."

The Magic of Numbers (p. 8)

. . . we shall leave to the antiquarians the difficult and delicate task of restoring the roses to the cheeks of mathematical mummies.

The Development of Mathematics
To Any Prospective Reader (p. v)

Just as "beauty is its own excuse for being," so mathematics needs no apology for existing.

The Queen of the Sciences
Chapter VI (p. 82)

Bill, Max

I am of the opinion that it is possible to develop an art largely on the basis of mathematical thinking.

Quoted in Eli Maor
To Infinity and Beyond: A Cultural History of the Infinite (p. 139)

Birns, Harold

One might risk establishing the following mathematical formula for bribery . . . namely OG = PLR × AEB: The opportunity for graft equals the plethora of legal requirements multiplied by the number of architects, engineers and builders.

New York Times
City Acts to Unify Inspection Rules
p. 43, column 8, October 2, 1963

Bolyai, John

Mathematical discoveries, like springtime violets in the woods, have their season which no human can hasten or retard.

Quoted in Israel Kleiner
Mathematics Teacher
Thinking the Unthinkable: The Story of Complex Numbers (with a Moral)
Volume 81, Number 7, October 1988 (p. 590)

Boutroux, E.

The mathematical laws presuppose a very complex elaboration. They are not known exclusively either a priori or a posteriori, but are a creation of the mind; and this creation is not an arbitrary one, but, owing to the mind's resources, takes place with reference to experience and in view of it. Sometimes the mind starts with intuitions which it freely creates; sometimes, by a process of elimination, it gathers up the axioms it regards as most suitable for producing a harmonious development, one that is both simple and fertile. The mathematics is a voluntary and intelligent adaptation of thought to things, it represents the forms that will allow of qualitative diversity being surmounted, the moulds into which reality must enter in order to become as intelligible as possible.

Natural Law in Science and Philosophy (p. 40)

Browder, Felix E.
MacLane, Saunders
The potential usefulness of a mathematical concept or technique in helping to advance scientific understanding has very little to do with what one can foresee before that concept or technique has appeared.

Quoted in Lynn Arthur Steen
Mathematics Today: Twelve Informal Essays
The Relevance of Mathematics (p. 348)

Burke, Edmund
It is from this absolute indifference and tranquillity of the mind, that mathematical speculations derive some of the most considerable advantages; because there is nothing to interest the imagination; because the judgment sits free and unbiased to examine the point. All proportions, every arrangement of quantity, is alike to the understanding, because the same truths result to it from all; from greater from lesser, from equality and inequality.

On the Sublime and Beautiful
Part 3, section 2

Butler, Nicholas Murray
The analytical geometry of Descartes and the calculus of Newton and Leibniz have expanded into the marvelous mathematical method—more daring in its speculations than anything that the history of philosophy records—of Lobachevsky and Riemann, Gauss and Sylvester. Indeed, mathematics, the indispensable tool of the sciences, defying the senses to follow its splendid flights, is demonstrating to-day, as it never has been demonstrated before, the supremacy of the pure reason.

The Meaning of Education and Other Essays and Addresses (p. 45)

Carlyle, Thomas
It is a mathematical fact that the casting of this pebble from my hand alters the centre of gravity of the universe.

Sartor Resartus
III

Carroll, Lewis
Again, it is often impossible for students to carry on accurate mathematical calculations in close contiguity to one another owing to their mutual interference, and a tendency to general conversation: consequently these processes require different rooms in which irrepressible conversationalists, who are found to occur in every branch of Society, might be carefully and permanently fixed.

It may be sufficient for the present to enumerate the following requisites: others might be added as funds permitted.

A. A very large room for calculating Greatest Common Measure. To this a small one might be attached for Least Common Multiple: this, however, might be dispensed with.

B. A piece of open ground for keeping Roots and practicing their extraction: it would be advisable to keep Square Roots by themselves, as their corners are apt to damage others.

C. A room for reducing Fractions to their Lowest Terms. This should be provided with a cellar for keeping the Lowest Terms when found, which might also be available to the general body of undergraduates, for the purpose of "keeping Terms".

D. A large room, which might be darkened, and fitted with a magic lantern, for the purpose of exhibiting Circulating Decimals in the act of circulation. This might also contain cupboards, fitted with glass-doors, for keeping the various Scales of Notation.

E. A narrow strip of ground, railed off and carefully leveled for investigating the properties of Asymptotes, and testing practically whether Parallel Lines meet or not . . .

> *The Complete Works of Lewis Carroll*
> The Offer of the Clarendon Trustees

Yet what are all such gaieties to me
Whose thought are full of indices and surds?

> *The Complete Works of Lewis Carroll*
> Phantasmagoria and other Poems

Casti, John L.

Mathematical modeling is about rules—the rules of reality. What distinguishes a mathematical model from, say, a poem, a song, a portrait or any other kind of "model", is that the mathematical model is an image or picture of reality painted with logical symbols instead of with words, sounds or watercolors. These symbols are then strung together in accordance with a set of rules expressed in a special language, the language of mathematics. A large part of the story told in the 800 pages or so comprising the two volumes of this work is about the grammar of this language. But a piece of the real world encoded into a set of mathematical rules (i.e. a model) is itself an abstraction drawn from the deeper realm of "the real thing." Based as it is upon a choice of what to observe and what to ignore, the real-world starting point of any mathematical model must necessarily throw away aspects of this "real thing" deemed irrelevant for the purposes of the model. So when trying to fathom the meaning of the title of this volume, I invite the reader to regard the word "rule" as either a noun or a verb—or even to switch back and forth between the two—according to taste.

> *Reality Rules*
> Volume I, Preface (p. vii)

Cayley, Arthur

As for everything else, so for a mathematical theory: beauty can be perceived but not explained.

Quoted by Eric Temple Bell in James R. Newman
The World of Mathematics
Volume I
Invariant Twins, Cayley and Sylvester (p. 341)

Chrystal, George

Every mathematical book that is worth reading must be read "backwards and forwards," if I may use the expression. I would modify Lagrange's advice a little and say, "Go on, but often return to strengthen your faith." When you come on a hard or dreary passage, pass it over; and come back to it after you have seen its importance or found the need for it further on.

Algebra
Part II
Preface (p. ix)

Colton, Walter

He that gives a portion of his time and talent to the investigation of mathematical truth will come to all other questions with a decided advantage over his opponents. He will be in argument what the ancient Romans were in the field: to them the day of battle was a day of comparative recreation because they were ever accustomed to exercise with arms much heavier than they fought; and reviews differed from a real battle in two respects: they encountered more fatigue, but the victory was bloodless.

Lacon (p. 178)

Colum, Padraic

An age being mathematical, these flowers
Of linear stalks and spheroid blooms were prized . . .

Quoted in Helen Plotz
Imagination's Other Place
Tulips

Comte, Auguste

We can now define Mathematical science with precision. It has for its object the *indirect* measurement of magnitudes, and it proposes *to determine magnitudes by each other, according to the precise relations which exist between them.*

The Positive Philosophy
Volume I
Book I, Chapter I (p. 38)

Mathematical Analysis is therefore the true rational basis of the whole system of our positive knowledge.

The Positive Philosophy
Volume I
Book I, Chapter I (p. 42)

It must be ever remembered that the true positive spirit first came forth from the pure sources of mathematical science; and it is only the mind that has imbibed it there, and which has been face to face with the lucid truths of geometry and mechanics, that can bring into full action its natural positively, and apply it in bringing the most complex studies into the reality of demonstration. No other discipline can fitly prepare the intellectual organ.

The Positive Philosophy
Volume I
Book III, Chapter I (p. 221)

Condon, Edward U.
In these days when so much emphasis is properly being placed on economy in government research operations, it is important to take advantage of the substantial savings that can be effected by substituting sound mathematical analysis for costly experimentation. In science as well as in business, it pays to stop and figure things out in advance.

The National Applied Mathematical Laboratories

Cooley, Hollis R.
Because mathematics has left its imprint upon so many aspects of present day civilization, its position in the modern world is a fundamental one, and a knowledge of mathematics is essential for a comprehensive understanding of current life and thought.

Introduction to Mathematics (p. 615)

Copernicus, Nicolaus
But if perchance there are certain "idle talkers" who take it upon themselves to pronounce judgment although wholly ignorant of mathematics, and if by shamelessly distorting the sense of some passage in Holy Writ to suit their purpose, they dare to reprehend and to attack my work; they worry me so little that I shall even scorn their judgments as foolhardy.

On the Revolutions of the Heavenly Spheres
Preface and Dedication to Pope Paul III

Courant, Richard
Since the seventeenth century, physical intuition has served as a vital source for mathematical problems and methods. Recent trends and fashions have, however, weakened the connection between

mathematics and physics; mathematicians, turning away from the roots of mathematics in intuition, have concentrated on refinement and emphasized the postulated side of mathematics, and at times have overlooked the unity of their science with physics and other fields. In many cases, physicists have ceased to appreciate the attitudes of mathematicians. This rift is unquestionably a serious threat to science as a whole; the broad stream of scientific development may split into smaller and smaller rivulets and dry out. It seems therefore important to direct our efforts toward reuniting divergent trends by classifying the common features and interconnections of many distinct and diverse scientific facts.

Methods of Mathematical Physics
Volume I (pp. v–vi)

For scholars and layman alike it is not philosophy but active experience in mathematics itself that alone can answer the question: What is mathematics?

What is Mathematics? (p. xix)

Courant, Richard
Robbins, Herbert

Mathematics as an expression of the human mind reflects the active will, the contemplative reason, and the desire for aesthetic perfection. Its basic elements are logic and intuition, analysis and construction, generality and individuality. Though different traditions may emphasize different aspects, it is only the interplay of these antithetic forces and the struggle for their synthesis that constitute the life, usefulness, and supreme value of mathematical science.

What is Mathematics? (p. xv)

Cournot, Augustin

. . . those skilled in mathematical analysis know that its object is not simply to calculate numbers, but that it is also employed to find the relations between magnitudes which cannot be expressed in numbers and between functions whose law is not capable of algebraic expression.

Researches into the Mathematical Principles of the Theory of Wealth
Preface (p. 3)

d'Abro, A.

Success has attended the efforts of mathematical physicists in so large a number of cases that, however marvelous it may appear, we can scarcely escape the conclusion that nature must be rational and susceptible to mathematical law.

The Evolution of Scientific Thought from Newton to Einstein (p. xii)

da Vinci, Leonardo

There is no certainty where one can neither apply any of the mathematical sciences nor any of those which are based upon the mathematical sciences.

The Notebooks of Leonardo da Vinci
Volume I
Mathematics (p. 634)

Dantzig, Tobias

Mathematical achievement shall be measured by standards which are peculiar to mathematics. These standards are independent of a crude reality of our senses. They are: freedom from logical contradictions, the generality of the laws governing the created form, the kinship which exists between this new form and those that have preceded it.

Number: The Language of Science (p. 231)

Darwin, Charles

Every new body of discovery is mathematical in form, because there is no other guidance we can have.

Quoted in Stanley Gudder
A Mathematical Journey (p. 171)

de Morgan, Augustus

The moving power of mathematical invention is not reasoning but imagination.

Quoted in Robert Percevel Graves
Life of Sir W.R. Hamilton (p. 219)

The greatest writers on mathematical subjects have a genius which saves them from their own slips, and guides them to true results through inaccurate expression, and sometime through absolute error.

The Differential and Integral Calculus (p. 691)

Dee, John

A marvelous neutrality have these things Mathematical, and also a strange participation between things supernatural, immortal, intellectual, simple, and indivisible, and things natural, mortal, sensible, compounded, and divisible.

Quoted in Stanley Gudder
A Mathematical Journey (p. 45)

Dieudonné, Jean

. . . in the flowering of a mathematical talent social environment has an important part to play.

Mathematics—The Music of Reason (p. 9)

Dirac, Paul Adrien Maurice
From now on there will be no physical treatise which is not primarily mathematical.

> Quoted in Walter R. Fuchs
> *Mathematics for the Modern Mind* (p. 25)

The steady progress of physics requires for its theoretical formulations a mathematics that gets continually more advanced.

> *Proceedings of the Royal Society*
> Quantised Singularities in the Electromagnetic Field
> Series A, Volume 133, Number 821, September 1, 1931 (p. 60)

Theoretical physicists accept the need for mathematical beauty as an act of faith . . . For example, the main reason why the theory of relativity is so universally accepted is its mathematical beauty.

> *From a Life of Physics*
> Lecture 2
> Methods in Theoretical Physics (p. 22)

Doob, J.L.
The basic difference between the roles of mathematical probability in 1946 and 1988 is that the subject is now accepted as mathematics whereas in 1946 to most mathematicians mathematical probability was to mathematics as black marketing to marketing; that is, probability was a source of interesting mathematics but examination of the background context was undesirable.

> Quoted in Peter Duren
> *A Century of Mathematics in America*
> Part II
> Commentary on Probability (p. 353)

Dyson, Freeman
One factor that has remained constant through all the twists and turns of the history of physical science is the decisive importance of the mathematical imagination.

> Quoted in Robert Osserman
> *Poetry of the Universe* (p. 125)

On being asked what he meant by the beauty of a mathematical theory of physics, Dirac replied that if the questioner was a mathematician then he did not need to be told, but were he not a mathematician then nothing would be able to convince him of it.

> Quoted in John D. Barrow
> *Theories of Everything* (p. 16)

The bottom line for mathematicians is that the architecture has to be right. In all the mathematics that I did, the essential point was to find the right architecture. It's like building a bridge. Once the main lines of the structure are right, then the details miraculously fit. The problem is the overall design.

The College Mathematics Journal
"Freeman Dyson: Mathematician, Physicist, and Writer":
Interview with Donald J. Albers
Volume 25, Number 1, January 1994

Farrar, John
. . . in mathematical science, and in it alone, man sees things precisely as God sees them, handles the very scale and compass with which the Creator planned and built the universe; . . .

As reported by Andrew P. Peabody in Florian Cajori
The Teaching and History of Mathematics in the United States
(p. 128, fn)

Fourier, [Jean Baptiste] Joseph
. . . mathematical analysis is as extensive as nature itself; it defines all perceptible relations, measures times, spaces, forces, temperatures; this difficult science is formed slowly, but it preserves every principle which it has once acquired; it grows and strengthens itself incessantly in the midst of the many variations and errors of the human mind.

Its chief attribute is clearness; it has no marks to express confused notions. It brings together phenomena the most diverse, and discovers the hidden analogies which unite them. If matter escapes us, as that of air and light, by its extreme tenuity, if bodies are placed far from us in the immensity of space, if man wishes to know the aspect of the heavens at successive epochs separated by a great number of centuries, if the actions of gravity and of heat are exerted in the interior of the earth at depths which will be always inaccessible, mathematical analysis can yet lay hold of the laws of these phenomena.

Analytical Theory of Heat
Preliminary Discourse

Profound study of nature is the most fertile source of mathematical discoveries. Not only has this study, in offering a determinate object to investigation, the advantage of excluding vague questions and calculations without issue; it is besides a sure method of forming analysis itself, and of discovering the elements which it concerns us to know, and which natural science ought always to preserve: these are the fundamental elements which are reproduced in all natural effects.

Analytical Theory of Heat
Preliminary Discourse

Hadamard, J.
The theory of integral equations, born yesterday, is already classical. It has been introduced in several university courses. There is no doubt—perhaps further improvements—that it will soon impose itself as of current use in mathematics. This is a rare piece of good fortune for a mathematical doctrine, for mathematical doctrines so often become museum exhibits.

Quoted in A. d'Abro
The Decline of Mechanism (p. 118)

Hankel, Hermann
If we compare a mathematical problem with an immense rock, whose interior we wish to penetrate, then the work of the Greek mathematicians appear to us like that of a robust stonecutter, who, with indefatigable perseverance, attempts to demolish the rock gradually from the outside by means of hammer and chisel; but the modern mathematician resembles an expert miner, who first constructs a few passages through the rock and then explodes it with a single blast, bringing to light its inner treasures.

Die Entwickelung der Mathematik in den letzten Jahrhunderten (p. 9)

Hardy, Godfrey Harold
We may say, roughly, that a mathematical idea is 'significant' if it can be connected, in a natural and illuminating way, with a large complex of other mathematical ideas.

A Mathematician's Apology
Chapter 11 (p. 89)

Heisenberg, Werner
The chain of cause and effect could be quantitatively verified only if the whole universe were considered as a single system—but then physics has vanished, and only a mathematical scheme remains. The partition of the world into observing and observed system prevents a sharp formulation of the law of cause and effect.

The Physical Principles of the Quantum Theory (p. 58)

Hermes, Hans
It may be permissible to compare mathematical research with the opening up of a mountain range. There will always be the people whose principal interest it will be to try their ability in advanced mountaineering. They will go for the most difficult summits. Others will see their aim in making the mountain range accessible as a whole, by building convenient roads along the valleys and across the passes. They will also reach the summits eventually, but mainly for the sake of the beautiful views, and, if possible, by cable car.

Quoted in Walter R. Fuchs
Mathematics for the Modern Mind (p. 18)

Hilbert, David
Mathematical science is in my opinion an indivisible whole, an organism whose vitality is conditioned upon the connection of its parts.

> *Bulletin of the American Mathematical Society*
> Mathematical Problems
> Volume 8, 2nd series, October 1901–July 1902 (p. 478)

Huxley, Thomas
Mathematical training is almost purely deductive. The mathematician starts with a few simple propositions, the proof of which is so obvious that they are called self-evident, and the rest of his work consists of subtle deductions from them.

> *Macmillan's Magazine*
> Scientific Education: Notes of an After Dinner Speech
> Volume XX, July 1869 (p. 182)

Jeans, Sir James Hopwood
The essential fact is simply that *all* the pictures which science now draws of nature, and which alone seem capable of according with observational fact, are *mathematical* pictures.

> *The Mysterious Universe*
> Into the Deep Waters (p. 150)

. . . it can hardly be disputed that nature and our conscious mathematical minds work according to the same laws.

> *The Mysterious Universe*
> Into the Deep Waters (p. 165)

Jevons, W. Stanley
In abstract mathematical theorems, the approximation to truth is perfect . . . In physical science, on the contrary, we treat of the least quantities which are perceptible.

> Quoted by J.W. Mellor
> *Higher Mathematics for Students of Chemistry and Physics* (p. 266)

Kasner, Edward
Newman, James R.
It was decided, after careful mathematical researches in the kindergarten, that the number of raindrops falling on New York in 24 hours, or even in a year or in a century, is much less than a google.
[10,000,000,000,000,000,000,000,000,000,000,000,000,000,000,000,000,000,
000,000,000,000,000,000,000,000,000,000,000,000,000,000,000]

> *Mathematics and the Imagination* (p. 20)

In purging mathematical philosophy of metaphysics, there has been . . . a real gain. No longer is mathematics to be looked upon as a key to the truth with a capital *T*. It may now be regarded as a woefully incomplete, though enormously useful, Baedeker in a mostly uncharted land. Some of the landmarks are fixed; some of the vast network of roads is made understandable; there are guideposts for the bewildered traveler.

Mathematics and the Imagination (p. 360)

Kepler, Johannes
If there is anything that can bind the heavenly mind of man to this dreary exile of our earthly home and can reconcile us with our fate so than one can enjoy living—then it is verily the enjoyment of the mathematical sciences and astronomy.

Quoted in H.E. Huntley
The Divine Proportion: a Study in Mathematical Beauty (p. 6)

Keyser, Cassius J.
The validity of mathematical propositions is independent of the actual world—the world of existing subject-matters—, is logically prior to it, and would remain unaffected were it to vanish from being.

The Pastures of Wonder
The Realm of Mathematics (p. 99)

Kline, Morris
The tantalizing and compelling pursuit of mathematical problems offers mental absorption, peace of mind amid endless challenges, repose in activity, battle without conflict, 'refuge from the goading urgency of contingent happenings', and the sort of beauty changeless mountains present to senses tried by the present-day kaleidoscope of events.

Mathematics in Western Culture (p. 470)

Lakatos, Imre
On the face of it there should be no disagreement about mathematical proof. Everybody looks enviously at the alleged unanimity of mathematicians; but in fact there is a considerable amount of controversy in mathematics. Pure mathematicians disown the proofs of applied mathematicians, while logicians in turn disavow those of pure mathematicians. Logicists disdain the proofs of formalists and some intuitionists dismiss with contempt the proofs of logicists and formalists.

Mathematics, Science and Epistemology
Volume 2 (p. 61)

Leibniz, Gottfried Wilhelm
All things in the whole wide world happen mathematically.

Quoted in Walter R. Fuchs
Mathematics for the Modern Mind (p. 14)

Lemoine, Émile
A mathematical truth is neither simple nor complicated in itself, it is.

Quoted in Eric T. Bell
Men of Mathematics (p. xv)

Locke, John
Mathematical proofs, like diamonds, are hard as well as clear, and will be touched with nothing but strict reasoning.

Quoted in Stanley Gudder
A Mathematical Journey (p. 20)

And thus all mathematical demonstrations, as well as first principles, must be received as native impressions on the mind; which I fear they will scarce allow them to be, who find it harder to demonstrate a proposition than assent to it when demonstrated. And few mathematicians will be forward to believe, that all the diagrams they have drawn were but copies of those innate characters which nature had engraven upon their minds.

An Essay Concerning Human Understanding
Book I, Chapter 1, Section 22

Merz, J.T.
In every case the awakening touch has been the mathematical spirit, the attempt to count, to measure, or to calculate. What to the poet or the seer may appear to be the very death of all his poetry and all his visions—the cold touch of the calculating mind,—this has proved to be the spell by which knowledge has been born, by which new sciences have been created, and hundreds of definite problems put before the minds and into the hands of diligent students. It is the geometrical figure, the dry algebraical formula, which transforms the vague reasoning of the philosopher into a tangible and manageable conception; which represents, though it does not explain, the things and processes of nature: this clothes the fruitful, but otherwise indefinite, ideas in such a form that the strict logical methods of thought can be applied, that the human mind can in its inner chamber evolve a train of reasoning the result of which corresponds to the phenomena of the outer world.

A History of European Thought in the Nineteenth Century
Volume 1 (p. 314)

Montague, William Pepperel
. . . that climax of rationalist aspiration in which even the arbitrary and existential constants of physics would be reduced to a crystalline precipitate of purely subsistential mathematical relations.

Philosophical Review
The Einstein Theory and a Possible Alternative
Volume 33, Number 193, March 1924 (p. 169)

Newman, M.H.A.

That mathematical theory is a lasting object to believe in few can doubt. Mathematical language is difficult but imperishable. I do not believe that any Greek scholar of to-day can understand the idiomatic undertones of Plato's dialogues, or the jokes of Aristophanes, as thoroughly as mathematicians can understand every shade of meaning in Archimedes' works.

Mathematical Gazette
What is Mathematics?
Volume 43, Number 345, October 1959 (p. 167)

Papert, Seymour

Mathematical work does not proceed along the narrow logical path of truth to truth to truth, but bravely or gropingly follows deviations through the surrounding marshland of propositions which are neither simply and wholly true nor simply and wholly false.

Mindstorms (p. 195)

Picard, Émile

I am not unaware of the difficulties of the task which I am undertaking. Activity in mathematical thinking today is such that it is perhaps presumptuous to attempt to sketch, in so vast an area, the present state of the science. The portrait, even if it is a good likeness, is fated, in parts at least, to become dated quickly. But that does not matter so long as I propose merely to be useful as a guide to those who wish to acquaint themselves with modern analysis and who fear that, alone, they may lose their way in the multiplicity of papers which fill the learned scientific periodicals.

Traite d'Analyse
Preface

Poincaré, Henri

The genesis of mathematical discovery is a problem which must intensely inspire the psychologist with the keenest interest. For this is the process in which the human mind seems to borrow least from the exterior world, in which it acts, or appears to act, only by itself and on itself, so that by studying the process of geometric thought we may hope to arrive at what is most essential in the human mind.

Science and Method
Mathematical Discovery (p. 47)

What, in fact, is mathematical discovery? It does not consist in making new combinations with mathematical entities that are already known. That can be done by any one, and the combinations that could be so formed would be infinite in number, and the greater part of them would be absolutely devoid of interest. Discovery consists precisely in not constructing useless combinations, but in constructing those that are useful, which are an infinitely small minority. Discovery is discernment, selection.

Science and Method
Mathematical Discovery (pp. 50–1)

It may appear surprising that sensibility should be introduced in connection with mathematical demonstrations, which, it would seem, can only interest the intellect. But not if we bear in mind the feeling of mathematical beauty, of the harmony of numbers and forms and of geometric elegance. It is a real aesthetic feeling that all true mathematicians recognize, and this is truly sensible.

Science and Method
Mathematical Discovery (p. 59)

Russell, Bertrand

Thus it would seem that wherever we infer from perceptions, it is only structure that we can validly infer; and structure is what can be expressed by mathematical logic, which includes mathematics.

The Analysis of Matter
Chapter XXIV (p. 254)

I wanted certainty in the kind of way in which people want religious faith. I thought that certainty is more likely to be found in mathematics than elsewhere. But I discovered that many mathematical demonstrations, which my teachers expected me to accept, were full of fallacies, and that, if certainty were indeed discoverable in mathematics, it would be in a new kind of mathematics, with more solid foundations than those that had hitherto been thought secure. But as the work proceeded, I was continually reminded of the fable about the elephant and the tortoise. Having constructed an elephant upon which the mathematical world could rest, I found the elephant tottering, and proceeded to construct a tortoise to keep the elephant from falling. But the tortoise was no more secure than the elephant, and after some twenty years of very arduous toil, I came to the conclusion that there was nothing more that I could do in the way of making mathematical knowledge indubitable.

Portraits from Memory and Other Essays
Reflections on My Eightieth Birthday (p. 54)

. . . none of the raw material of the world has smooth logical properties, but whatever appears to have such properties is constructed artificially to have them.

The Principles of Mathematics
Preface (p. xi)

. . . logic is concerned with the real world just as truly as zoology . . .
Introduction to Mathematical Philosophy (p. 169)

Sarton, George
A mathematical congress of to-day reminds one of the Tower of Babel, for few men can follow profitably the discussions of sections other than their own, and even there they are sometimes made to feel like strangers.
The Study of the History of Mathematics (p. 14)

Schubert, Hermann
The intrinsic character of mathematical research and knowledge is based essentially on three properties: first, on its conservative attitude towards the old truths and discoveries of mathematics; secondly, on its progressive mode of development, due to the incessant acquisition of new knowledge on the basis of the old; and thirdly, on its self-sufficiency and its consequent absolute independence.
Mathematical Essays and Recreations (p. 27)

. . . the three positive characteristics that distinguish mathematical knowledge from other knowledge . . . may be briefly expressed as follows; first, mathematical knowledge bears more distinctly the imprint of truth on all its results than any other kind of knowledge; secondly, it is always a sure preliminary step to the attainment of other correct knowledge; thirdly, it has no need of other knowledge.
Mathematical Essays and Recreations (p. 35)

Seneca
The mathematical is, so to speak, a superficial science; it builds on a borrowed site, and the principles by aid of which it proceeds, are not its own . . .

Quoted in
Edinburgh Review
Volume 52
January 1836 (p. 221)

Siegel, Eli
Within all conflagrations mathematical things are related.

Damned Welcome
Aesthetic Realism Maxims
Part I, #356 (p. 78)

Simmons, G.F.
Mathematical rigor is like clothing; in its style it ought to suit the occasion, and it diminishes comfort and restricts freedom of movement if it is either too loose or too tight.

The Mathematical Intelligencer
Filler
Volume 13, Number 1, Winter 1991 (p. 69)

Swann, W.F.G.
. . . the mathematical physicist [is] one who obtains much prestige from the physicists because they are impressed with the amount of mathematics he knows, and much prestige from the mathematicians, because they are impressed with the amount of physics he knows.

The Architecture of the Universe (p. 14)

Swift, Jonathan
At last we entered the palace, and proceeded into the chamber of presence, where I saw the king seated on his throne, attended on each side by persons of prime quality. Before the throne was a large table filled with globes and spheres, and mathematical instruments of all kinds.

Gulliver's Travels
Part III, Chapter II

My dinner was brought, and four persons of quality, whom I remembered to have seen very near the king's person, did me the honor to dine with me. We had two courses of three dishes each. In the first course there was a shoulder of mutton, cut into an equilateral triangle, a piece of beef into a rhomboids, and a pudding into a cycloid. The second course was two ducks, trussed up into the form of fiddles; sausages and puddings resembling flutes and hautboys, and a breast of veal in the shape of a harp. The servants cut our bread into cones, cylinders, parallelograms, and several other mathematical figures.

Gulliver's Travels
Part III, Chapter II

The knowledge I had in mathematicks gave me great assistance in acquiring their phraseology, which depended much upon that science and musick; and in the latter I was not unskilled. Their ideas are perpetually conversant in lines and figures. If they would, for example, praise the beauty of a woman, or any other animal, they describe it by rhombs, circles, parallelograms, ellipses, and other geometrical terms, or by words of art drawn from music, needless here to repeat. I observed in the king's kitchen all sorts of mathematical and musical instruments, after the figures of which they cut up the joynts that were served to his Majesty's table.

Gulliver's Travels
Part III, Chapter II

Sylvester, James Joseph
Number, place, and combination . . . the three intersecting but distinct spheres of thought to which all mathematical ideas admit of being referred.

Philosophical Magazine
Volume 24, 1844 (p. 285)

There are three ruling ideas, three so to say, spheres of thought, which pervade the whole body of mathematical science, to some one or other of which, or to two or all three of them combined, every mathematical truth admits of being referred; these are the three cardinal notions, of Number, Space and Order.

Arithmetic has for its object the properties of number in the abstract. In algebra, viewed as a science of operations, order is the predominating idea. The business of geometry is with the evolution of the properties of space, or of bodies viewed as existing in space . . .

The Collected Mathematical Papers of James Joseph Sylvester
Volume II
A Probationary Lecture on Geometry, Delivered before the
Gresham Committee and the Members of the
Common Council of the City of London, 4 December, 1854 (p. 5)

. . . we are told that "Mathematics is that study which knows nothing of observation, nothing of induction, nothing of causation." I think no statement could have been made more opposite to the undoubted facts of the case, that mathematical analysis is constantly invoking the aid of new principles, new ideas, and new methods, not capable of being defined by any form of words, but springing direct from the inherent powers and activity of the human mind, and from continually renewed introspection of that inner world of thought of which the phenomena are as varied and require as close attention to discern as those of the outer physical (to which the inner one in each individual man may, I think, be conceived to stand in somewhat the same general relation of correspondence as a shadow to the object from which it is projected, or as the hollow palm of one hand to the closed fist which it grasps of the other), that it is unceasingly calling forth the faculties of observation and comparison, that one of its principal weapons is induction, that it has frequent recourse to experimental trial and verification, and that it affords a boundless scope for the exercise of the highest efforts of imagination and invention.

The Collected Mathematical Papers of James Joseph Sylvester
Volume II
Presidential Address to Section 'A' of the British Association (p. 654)

Thompson, Sir D'Arcy Wentworth
The harmony of the world is made manifest in Form and Number, and the heart and soul and all the poetry of Natural Philosophy are embodied in the concept of mathematical beauty.

On Growth and Form
Volume I
Chapter 10

Thoreau, Henry David
The most distinct and beautiful statements of any truth must take at last the mathematical form. We might so simplify the rules of moral philosophy, as well as of arithmetic, that one formula would express them both.

A Week on the Concord and Merrimac Rivers
Friday (p. 323)

Unknown
A mathematical theory is not to be considered complete until you have made it so clear that you can explain it to the first man whom you meet on the street.

Bulletin of the American Mathematical Society
Mathematical Problems
Volume 8, 2nd series, October 1901–July 1902 (p. 438)

von Neumann, John
After all, classical mathematics was producing results which were both elegant and useful, and, even though one could never again be absolutely certain of its reliability, it stood on at least as sound a foundation as, for example, the existence of the electron. Hence, if one was willing to accept the sciences, one might as well accept the classical system of mathematics.

Collected Works
Volume I, The Works of the Mind
The Mathematician (p. 6)

. . . much of the best mathematical inspiration comes from experience and that it is hardly possible to believe in the existence of an absolute, immutable concept of mathematical rigor, dissociated from all human experience.

Collected Works
Volume I, The Works of the Mind
The Mathematician (p. 6)

. . . at a great distance from its empirical source, or after much "abstract" inbreeding, a mathematical subject is in danger of degeneration. At the inception the style is usually classical; when it shows signs of becoming baroque, then the danger signal is up . . . In any event, whenever this stage is reached, the only remedy seems to me to be the rejuvenating return to the source: the reinjection of more or less directly empirical ideas. I am convinced that this was a necessary condition to conserve freshness and vitality of the subject and that this will remain equally true in the future.

Collected Works
Volume I, The Works of the Mind
The Mathematician (p. 6 and p. 9)

Weyl, Hermann

The stringent precision attainable for mathematical thought has led many authors to a mode of writing which must give the reader the impression of being shut up in a brightly illuminated cell where every detail sticks out with the same dazzling clarity, but without relief. I prefer the open landscape under the clear sky with its depth of perspective, where the wealth of sharply defined nearby details gradually fades away towards the horizon.

The Classical Groups; Their Invariants and Representations
Preface

While in other fields brief allusions are met by ready understanding, this is unfortunately seldom the case with mathematical ideas.

Quoted in K. Chandrasekharan (editor)
Hermann Weyl (p. 84)

You should not expect me to describe the mathematical way of thinking much more clearly than one can describe, say, the democratic way of life.

Quoted in K. Chandrasekharan (editor)
Hermann Weyl (p. 84)

Whitehead, Alfred North

In order that a mathematical science of any importance may be founded upon conventional definitions, the entities created by them must have properties which bear some affinity to the properties of existing things.

A Treatise on Universal Algebra
Preface (p. vii)

Williams, Horatio B.

Once a statement is cast into mathematical form it may be manipulated in accordance with these rules and every configuration of the symbols will represent facts in harmony with and dependent on those contained in the original statement. Now this comes very close to what we conceive the action of the brain structures to be in performing intellectual acts with the symbols of ordinary language. In a sense, therefore, the mathematician has been able to perfect a device through which a part of the labor of logical thought is carried on outside the central nervous system with only that supervision which is requisite to manipulate the symbols in accordance with the rules.

Bulletin of the American Mathematical Society
Mathematics and the Biological Sciences
Volume 38, May–June 1927 (p. 291)

MATHEMATICAL WRITING

Pétard, H. [Pondiczery, E.S.]
Since authors seldom, if ever, say what they mean, the following glossary is offered to neophytes in mathematical research to help them understand the language that surrounds the formulas . . .

ANALOGUE. This is an a. of: I have to have *some* excuse for publishing it.

APPLICATIONS. This is of interest in a.: I have to have *some* excuse for publishing it.

COMPLETE. The proof is now c.: I can't finish it.

DETAILS. I cannot follow the d. of X's proof: It's wrong. We omit the d.: I can't do it.

DIFFICULT. This problem is d.: I don't know the answer. (Cf. Trivial.)

GENERALITY. Without loss of g.: I have done an easy special case.

IDEAS. To fix the I.: To consider the only case I can do.

INGENIOUS. X's proof is I.: I understand it.

INTEREST. It may be of I. I have to have *some* excuse for publishing it.

INTERESTING. X's paper is I.: I don't understand it.

KNOWN. This is a k. result but I reproduce the proof for the convenience of the reader: My paper isn't long enough.

LANGUAGE. *Par abus de l.*: In the terminology used by other authors. (Cf. Notation.)

NATURAL. It is n. to begin with the following considerations: We have to start somewhere.

NEW. This was proved by X but the following n. proof may present points of interest: I can't understand X.

NOTATION. To simplify the n.: It is too much trouble to change now.

OBSERVED. It will be o. that: I hope you have not noticed that.

OBVIOUS. It is o.: I can't prove it.

READER. The details may be left to the r.: I can't do it.

REFEREE. I wish to thank the r. for his suggestions: I loused it up.

STRAIGHTFORWARD. By a s. computation: I lost my notes.

TRIVIAL. This problem is t.: I know the answer (Cf. Difficult.)

WELL-KNOWN. The result is w.: I can't find the reference.

The American Mathematical Monthly
A Brief Dictionary of Phrases Used in Mathematical Writing
Volume 73, Number 2, February 1966 (pp. 196–7)

Unknown

BRIEFLY: I'm running out of time, so I'll just write and talk faster.

BRUTE FORCE (AND IGNORANCE): Four special cases, three counting arguments, two long inductions, "and a partridge in a pair tree."

BY A PREVIOUS THEOREM: I don't remember how it goes (come to think of it I'm not really sure we did this at all), but if I stated it right (or at all), then the rest of this follows.

CANONICAL FORM: 4 out of 5 mathematicians surveyed recommended this as the final form for their students who choose to finish.

CHECK or CHECK FOR YOURSELF: This is the boring part of the proof, so you can do it on your own time.

CLEARLY: I don't want to write down all the "in-between" steps.

CLEARLY: Obscurely.

DIDN'T JONES DO SOME SIMILAR STUFF A FEW YEARS AGO?: I know where you copied.

ELEGANT PROOF: Requires no previous knowledge of the subject matter and is less than ten lines long.

GO ASK PROFESSOR SMITH OVER THERE: In his face.

HAVE YOU HAD MANY STUDENTS?: Do you have any social diseases?

HE'S ONE OF THE GREAT LIVING MATHEMATICIANS: He's written 5 papers and I've read 2 of them.

HINT: The hardest of several possible ways to do a proof.

HOW DO YOU RECONCILE YOUR THEOREM WITH THIS EXAMPLE?: You're dead.

I DON'T UNDERSTAND THAT STEP: You goofed.

I READ ONE OF YOUR PAPERS: I wrapped fish with one of your papers.

I THOUGHT ABOUT THAT PROBLEM 20 YEARS AGO, BUT I FORGET THE ANSWER: I'm more famous than you are.

I'LL SEND YOU SOME OF MY PAPERS: Drop dead.

IT CAN EASILY BE SHOWN: Even you, in your finite wisdom, should be able to prove this without me holding your hand.

IT'S TRUE. READ MY BOOK: I don't know.

I'VE HEARD SO MUCH ABOUT YOU: Stalling a minute may give me time to recall who you are.

LET'S MAKE AN APPOINTMENT AND TALK ABOUT IT: In my face.

LET'S TALK THROUGH IT: I don't want to write it on the board lest I make a mistake.

LET'S WRITE A JOINT PAPER ON IT: We're equally famous but I'm lazy.

LOOK IT UP IN DUNFORD AND SCHWARTZ: In your face.

OBVIOUSLY: I hope you weren't sleeping when we discussed this earlier, because I refuse to repeat it.

PROCEED FORMALLY: Manipulate symbols by the rules without any hint of their true meaning (popular in pure math courses).

PROOF OMITTED: Trust me.

QUANTIFY: I can't find anything wrong with your proof except that it won't work if x is a moon of Jupiter (popular in applied math courses).

RECALL: I shouldn't have to tell you this, but for those of you who erase your memory tapes after every test.

SEND ME YOUR PREPRINTS: Please go away.

SEND ME YOUR REPRINTS: Please stay away.

SIMILARLY: At least one line of the proof of this case is the same as before.

SKETCH OF A PROOF: I couldn't verify all the details, so I'll break it down into the parts I couldn't prove.

SOFT PROOF: One third less filling (of the page) than your regular proof, but it requires two extra years of course work just to understand the terms.

TFAE (The Following Are Equivalent): If I say this it means that, and if I say that it means the other thing, and if I say the other thing.

THAT PROBLEM IS INTRACTABLE: I can't do that problem so neither can you.

THAT'S THE MOST INSIGHTFUL QUESTION I EVER HEARD: You're more famous than I am.

THIS IS A CALCULATION: Let's all forget the proof.

THIS IS OBVIOUS: You forget the proof.

THIS IS TRIVIAL: I forget the proof.

TRIVIAL: If I have to show you how to do this, you're in the wrong class.

TWO LINE PROOF: I'll leave out everything but the conclusion, you can't question 'em if you can't see 'em.

WHAT ARE SOME APPLICATIONS OF YOUR THEOREM?: What is your theorem?

WHAT WAS YOUR THESIS ABOUT?: Are you still polishing your thesis?

WHERE DO YOU TEACH?: Do you have a job?

WHO WAS YOUR ADVISOR?: What rock did you crawl out from under?

WLOG (Without Loss Of Generality): I'm not about to do all the possible cases, so I'll do one and let you figure out the rest.

YOUR TALK WAS VERY INTERESTING: I can't think of anything to say about your talk.

YOUR THEOREM CONTRADICTS MY THEOREM: I'm dead.

Source unknown

O'Brien, Katharine
Said an upside-down A to an inside-out E,
"*Universal*'s the epithet measuring me.
   Your scope is so small
   Compared with *For all*—
There is no more than of form of *To be*."

The Mathematical Magazine
∀ and ∃ (p. 41)
Volume 55, Number 1, January 1982

MATHEMATICIAN

Adams, Henry
Mathematicians assume the right to choose, within the limits of logical contradiction, what path they please in reaching their results.
A Letter to American Teachers of History (p. v)

Mathematicians practice absolute freedom.
A Letter to American Teachers of History (p. 169)

He supposed that, except musicians, every one thought Beethoven a bore, as every one except mathematicians thought mathematics a bore.
The Education of Henry Adams
Chapter V (p. 80)

Adler, Alfred
Each generation has its few great mathematicians, and mathematics would not even notice the absence of the others. They are useful as teachers, and their research harms no one, but it is of no importance at all. A mathematician is great or he is nothing.
New Yorker Magazine
Mathematics and Creativity
February 19, 1972

In the company of friends, writers can discuss their books, economists the state of the economy, lawyers their latest cases, and businessmen their latest acquisitions, but mathematicians cannot discuss their mathematics at all. And the more profound their work, the less understandable it is.
New Yorker Magazine
Mathematics and Creativity
February 19, 1972

The mathematical life of a mathematician is short. Work rarely improves after the age of twenty-five or thirty. If little has been accomplished by then, little will ever be accomplished.
New Yorker Magazine
Mathematics and Creativity
February 19, 1972

. . . the mathematician learns early to accept no fact, to believe no statement, however apparently reasonable or obvious or trivial, until it has been proved, rigorously and totally by a series of steps proceeding from universally accepted first principles.

New Yorker Magazine
Mathematics and Creativity
February 19, 1972

Perhaps mathematicians, lacking the imagination to appreciate the scope and sophistication of the outside world, confuse minor success with real achievement and are satisfied with it. Then, too, they seldom recognize failure when they are confronted with it; rather, they tend to think of it as simply one more betrayal by a society that usually patronizes them while elevating armies of patently inferior claimants. In the academic world, on the other hand, mathematicians often enjoy rewards that they do not merit. They are engulfed by admirers from the department of philosophy and the social sciences . . .

New Yorker Magazine
Mathematics and Creativity
February 19, 1972

Barrow, Isaac
An accomplished mathematician, i.e. a most wretched orator.

Mathematical Lectures (p. 32)

It may be observed of mathematicians that they only meddle with such things as are certain, passing by those that are doubtful and unknown. They profess not to know all things, neither do they affect to speak of all things. What they know to be true, and can make good by invincible arguments, that they publish and insert among their theorems. Of other things they are silent and pass no judgment at all, choosing rather to acknowledge their ignorance, than affirm anything rashly.

Mathematical Lectures (p. 64)

Because Mathematicians frequently make use of Time, they ought to have a distinct idea of the meaning of that Word, otherwise they are Quacks . . .

Quoted by Paul Davies
About Time (p. 183)

Bell, Eric T.
Fools have always been governed by fools and doubtless always will be, but not all scientists and mathematicians are yet fools.

Mathematics: Queen and Servant of Science
Points of View (p. 13)

If mathematics is indeed the science of self-evident things, mathematicians are a phenomenally stupid lot to waste the tons of good paper they do in proving the fact.

Mathematics: Queen and Servant of Science
Mathematical Truth (p. 20)

Mathematicians are not, as a rule, credulous; their clients almost invariably are.

Mathematics: Queen and Servant of Science
Choice and Chance (p. 381)

Experience has taught most mathematicians that much that looks solid and satisfactory to one mathematical generation stands a fair chance of dissolving into cobwebs under the steadier scrutiny of the next . . .

Quoted in Morris Kline
Mathematics: The Loss of Certainty (p. 257)

The toughminded suggest that the theory of the infinite elaborated by the great mathematicians of the Nineteenth and Twentieth Centuries, without which mathematical analysis as it is actually used today is impossible, has been committing suicide in an unnecessarily prolonged and complicated manner for the past half century.

Debunking Science

The mathematician is a much rarer character in fiction than his cousin the scientist, and when he does appear in the pages of a novel or on the screen he is only too apt to be a slovenly dreamer totally devoid of common sense—comic relief.

Men of Mathematics (p. 8)

Mathematicians may safely be left to follow their own bent as their contributions to this age of science. What they did in the past century is enough for a vast region of science and technology as they exist today; what mathematicians as professionals are interested in today will, if there is any continuity at all in scientific and industrial history, be the indispensable framework of the science and technology of tomorrow.

The Queen of the Sciences
Chapter VI (p. 83)

That wretched monosyllable "all" has caused mathematicians more trouble than all the rest of the dictionary.

The Queen of the Sciences
Chapter X (p. 134)

Bernoulli, Daniel
. . . there is no philosophy which is not founded upon knowledge of the phenomena, but to get any profit from this knowledge it is absolutely necessary to be a mathematician.

Quoted in C. Truesdell
Essays in the History of Mechanics (p. 318)

Black, Max
. . . a result once generally accepted by mathematicians is seldom retracted, and then only with great pangs.

The Nature of Mathematics (p. 169)

Boltzmann, Ludwig Edward
A mathematician will recognize Cauchy, Gauss, Jacobi, or Helmholtz after reading a few pages, just as musicians recognize, from the first few bars, Mozart, Beethoven, or Schubert.

Quoted in Arthur Koestler
The Act of Creation (p. 265)

Bonaparte, Napoleon
Lagrange is the lofty pyramid of the mathematical sciences.

Quoted in E.T. Bell
Men of Mathematics (p. 153)

Bourbaki, Nicolas
For twenty-five centuries mathematicians have been in the habit of correcting their errors—and seeing their science enriched rather than impoverished thereby. This gives them the right to contemplate the future with serenity.

Quoted in Lucienne Felix
The Modern Aspect of Mathematics (p. 33)

Many mathematicians take up quarters in a corner of the domain of mathematics, which they do not intend to leave; not only do they ignore almost completely what does not concern their special field, but they are unable to understand the language and the terminology used by colleagues who are working in a corner remote from their own. Even among those who have the widest training, there are none who do not feel lost in certain regions of the immense world of mathematics; those who, like Poincaré or Hilbert, put the seal of their genius on almost every domain, constitute a very great exception even among the men of greatest accomplishment.

Quoted in Morris Kline
Mathematics: The Loss of Certainty (p. 284)

There is no sharply drawn line between those contradictions which occur in the daily work of every mathematician, beginner or master of his craft, as a result of more or less easily detected mistakes, and the major paradoxes which provide food for logical thought for decades and sometimes centuries.

Quoted in Bryan H. Bunch
Mathematical Fallacies and Paradoxes (p. 38)

Boyd, William Andrew Murray

The natural world is full of irregularity and random alteration, but in the antiseptic, dust-free, shadowless, brightly lit, abstract realm of the mathematicians they like their cabbages spherical, please.

Brazzaville Beach
Cabbages Are Not Spheres (p. 86)

Bush, Vannevar

A mathematician is not a man who can readily manipulate figures; often he cannot. He is not even a man who can readily perform the transformations of equations by the use of calculus. He is primarily an individual who is skilled in the use of symbolic logic on a high plane, and especially he is a man of intuitive judgment in the choice of the manipulative processes he employs.

Endless Horizons (p. 27)

Carmichael, R.D.

. . . the mathematician lives long and lives young; the wings of his soul do not early drop off.

The Logic of Discovery (p. 272)

Chesterton, Gilbert Keith

Poets do not go mad; but chess-players do. Mathematicians go mad, and cashiers; but creative artists very seldom.

Orthodoxy
The Maniac (p. 27)

Coleridge, Samuel Taylor

. . . from the time of Kepler to that of Newton, and from Newton to Hartley, not only all things in external nature, but the subtlest mysteries of life and organization, and even of the intellect and moral being, were conjured within the magic circle of mathematical formulae.

The Theory of Life (p. 355)

Courant, Richard

It becomes the urgent duty of mathematicians, therefore, to mediate about the essence of mathematics, its motivations and goals and the ideas that must bind divergent interests together.

Scientific American
Mathematics in the Modern World
Volume 211, Number 3, September 1964 (p. 42)

Copernicus, Nicolaus
Mathematics is written for mathematicians . . .

> Preface and Dedication to Pope Paul III
> *On the Revolutions of the Heavenly Spheres*

Crick, Francis Harry Compton
. . . in my experience most mathematicians are intellectually lazy and especially dislike reading experimental papers.

> *What Mad Pursuit* (p. 136)

da Vinci, Leonardo
Let no one read me who is not a mathematician in my beginnings.

> *The Notebooks of Leonardo da Vinci*
> Volume I
> Philosophy (p. 90)

In order to make trial of anyone and see whether he has a true judgment as to the nature of weights, ask him at what point one ought to cut one of the two equal arms of the balance so as to cause the part cut off, attached to the extremity to its remainder, to form with precision a counterpoise to the opposite arm. The thing is never possible, and if he gives you the position it is clear that he is a poor mathematician.

> *The Notebooks of Leonardo da Vinci*
> Volume II
> Experiments (p. 156)

Dantzig, Tobias
The mathematician may be compared to a designer of garments, who is utterly oblivious of the creatures who his garments may fit. To be sure, his art originated in the necessity for clothing such creatures, but this was long ago; to this day a shape will occasionally appear which will fit into the garment as if the garment had been made for it. Then there is no end of surprise and of delight.

> *Number: The Language of Science* (pp. 231–2)

Darwin, Charles
A mathematician is a blind man in a dark room looking for a black hat which isn't there.

> Attributed
> In Fort Tomlinson
> *American Mathematical Monthly*
> Mathematics and the Sciences
> Volume 47, November 1940 (p. 606)

Davis, Philip J.
Hersh, Reuben
The ideal mathematician's work is intelligible only to a small group of specialists, numbering a few dozen or at most a few hundred. This group

has existed for only a few decades, and there is every possibility that it may become extinct in another few decades. However, the mathematician regards his work as part of the very structure of the world, containing truths that are valid forever, from the beginning of time, even in the most remote corner of the universe.

The Mathematical Experience
The Ideal Mathematician (p. 38)

Day-Lewis, Cecil
They say that a mathematician
Once fell to such a passion
For x and y, he locked
His door to keep outside
Whatever might distract
Him from his heavenly bride:
And presently died
In the keenest of blisses
With a dozen untasted dishes
Outside his door.

Collected Poems 1929–1933
Transitional Poem
Part II, 16

de Fontenelle, Bernard
Mathematicians are like lovers . . . Grant a mathematician the least principle, and he will draw from it a consequence which you must also grant him, and from this consequence another.

Quoted in Eric T. Bell
Men of Mathematics (p. xv)

. . . it was the business of the Sorbonne to discuss; of the Pope to decide; and of the mathematician to go to heaven in a perpendicular line.

Quoted in Dugald Stewart
The Collected Works of Dugald Stewart
Volume IV (p. 203)

de Morgan, Augustus
We know that mathematicians care no more for logic than logicians for mathematics. The two eyes of exact science are mathematics and logic: the mathematical sect puts out the logical eye, the logical sect puts out the mathematical eye, each believing that it can see better with one eye than with two.

Quoted in Florian Cajori
The History of Mathematics (p. 316)

Dieudonné, Jean
. . . there is no criterion for appreciation which does not vary from
one epoch to another and from one mathematician to another . . .
These divergences in taste recall the quarrels aroused by works of art,
and it is a fact that mathematicians often discuss among themselves
whether a theorem is more or less "beautiful". This never fails to surprise
practitioners of other sciences: for them the sole criterion is the "truth" of
a theory or formula; . . . Other criteria are therefore necessary to evaluate
mathematical work, and these are unavoidably subjective, a fact which
makes some people say that mathematics is much more an art than a
science.

Mathematics—The Music of Reason (p. 28)

Egrafov, M.
If you ask mathematicians what they do, you always get the same
answer. They think. They think about difficult and unusual problems.
They do not think about ordinary problems: they just write down the
answers.

Mathematics Magazine
Volume 65, Number 5, December 1992 (p. 301)

Eliot, George
Every man who is not a monster, a mathematician, or a mad philosopher,
is the slave of some woman or other.

The Works of George Eliot
Scenes of Clerical Life
Volume I
The Sad Fortunes of the Rev. Amos Barton
Chapter iv (p. 61)

Esar, Evan
[Mathematician] A scientist who can figure out anything except such
simple things as squaring the circle and trisecting an angle.

Esar's Comic Dictionary

Euler, Leonhard
. . . the great German mathematician Leonhard Euler confronted the
eminent French scholar and atheist Denis Diderot with a spurious
mathematical proof for the existence of God. Euler, it seems, accepted
an invitation to meet Diderot, who at the time was in attendance at the
royal court of the Russian Czar. On the day or his arrival, the story goes,
Euler strode up to Diderot and proclaimed:

$$\text{"Monsieur, } (a + bn)/n = X, \text{ donc Dieu existe; répondez!"}$$
$$[\text{"Sir, } (a + bn)/n = X, \text{ therefore, God exists; respond!"}]$$

In the past, the French scholar had eloquently and forcefully refuted many a clever philosophical argument for the existence of God, but at this moment, at a loss to comprehend the meaning of this mathematical equation, Diderot was intimidated into silence.

Quoted in Michael Guillen
Bridges to Infinity (p. 1)

Flammarion, Camille
Mathematicians, whose tempers are generally intolerable, are perhaps psychologically excusable, for the constant tension of their mind is, perhaps, the cause of their bad digestion and their state of hypochondria.

Popular Astronomy (pp. 346–7, fn)

Glaisher, J.W.
The mathematician requires tact and good taste at every step of his work, and he has to learn to trust to his own instinct to distinguish between what is really worthy of his efforts and what is not.

Quoted in H. Eves
Mathematical Circles Squared
1089°

Goethe, Johann Wolfgang von
Mathematicians are a kind of Frenchman. They translate into their own language whatever is said to them and forthwith the thing is utterly changed.

Goethe's Poems and Aphorisms (p. 199)

. . . the mathematician is only complete in so far as he feels within himself the *beauty* of the true.

Quoted in Oswald Spengler
The Decline of the West
Volume I
Chapter II, section iv (p. 61)

Graham, L.A.
A mathematician named Ray
Says extraction of cubes is child's play.
   You don't need equations
   Or long calculations
Just hot water to run on the tray.

Ingenious Mathematical Problems and Methods
Mathematical Nursery Rhyme No. 14

Guillen, Michael
[Mathematicians] might actually be looking at life with a most trenchant sense—one that perceives things the other five senses cannot.

Bridges to Infinity (p. 7)

Hammersley, J.
People do acquire a little brief authority by equipping themselves with jargon: they can pontificate and air a superficial expertise. But what we should ask of educated mathematicians is not what they can speechify about, nor even what they know about the existing corpus of mathematical knowledge, but rather what can they now do with their learning and whether they can actually solve mathematical problems arising in practice. In short, we look for deeds not words.

In Institute of Mathematics and Its Applications
Bulletin
On the Enfeeblement of Mathematical Skills by 'Modern Mathematics' and
by Similar Soft Intellectual Truths in Schools and Universities
Volume 4, Number 2, October 1968

Hardy, Godfrey Harold
It is a melancholy experience for a professional mathematician to find himself writing about mathematics. The function of a mathematician is to do something, to prove new theorems, to add to mathematics, and not to talk about what he or other mathematicians have done . . . there is no scorn more profound, or on the whole more justifiable, than that of the men who make for the men who explain. Exposition, criticism, appreciation, is work for second-rate minds.

A Mathematician's Apology
Chapter 1 (p. 61)

If a man is in any sense a real mathematician, then it is a hundred to one that his mathematics will be better than anything else he can do, and that he would be silly if he surrendered any decent opportunity of exercising his one talent in order to do undistinguished work in other fields.

A Mathematician's Apology
Chapter 3 (p. 70)

No mathematician should ever allow himself to forget that that mathematics, more than any other art or science, is a young man's game.

A Mathematician's Apology
Chapter 4 (p. 70)

Archimedes will be remembered when Aeschylus is forgotten, because languages die and mathematical ideas do not. 'Immortality' may be a silly word, but probably a mathematician has the best chance of whatever it may mean.

A Mathematician's Apology
Chapter 8 (p. 81)

A MATHEMATICIAN, like a painter or a poet, is a maker of patterns. If his patterns are more permanent than theirs it is because they are made with *ideas*. A painter makes patterns with shapes and colours, a poet with words . . . A mathematician, on the other hand, has no material to work with but ideas, and so his patterns are likely to last longer . . .

A Mathematician's Apology
Chapter 10 (p. 84)

. . . *reductio ad absurdum* . . . is one of a mathematician's finest weapons. It is a far finer gambit than any chess gambit: a chess player may offer the sacrifice of a pawn or even a piece, but the mathematician offers *the game*.

A Mathematician's Apology
Chapter 12 (p. 94)

Heaviside, Oliver

Even Cambridge mathematicians deserve justice.

Quoted in Harold Jeffreys and Bertha Swirles
Methods of Mathematical Physics (p. 228)

But it is perhaps too much to expect a man to be both the prince of experimentalists and a competent mathematician.

Electromagnetic Theory
Volume I
Chapter I (p. 14)

Hughes, Richard

Science, being human enquiry, can hear no answer except an answer couched somehow in human tones. Primitive man stood in the mountains and shouted against a cliff; the echo brought back his own voice and he believed it a disembodied spirit. The scientist of today stands counting out loud in the face of the unknown. Numbers come back to him—and he believes in the Great Mathematician.

Quoted in Jefferson Hane Weaver
The World of Physics
Volume III (p. 597)

Huxley, Thomas H.

The Mathematician deals with two properties of objects only, number and extension, and all the inductions he wants have been formed and finished ages ago. He is now occupied with nothing but deductions and verification.

Lay Sermons, Addresses and Reviews (p. 87)

Inge, William Ralph

A mathematician, it has been suggested, might pray to x^n.

A Rustic Moralist (p. 7)

Jeans, Sir James Hopwood
. . . we may say that we have already considered with disfavor the possibility of the universe having been planned by a biologist or an engineer; from the intrinsic evidence of his creation, the Great Architect of the Universe now begins to appear as a pure mathematician.

The Mysterious Universe
Into the Deep Waters (p. 165)

Kasner, Edward
The great mathematicians have acted on the principle "Divinez avant de démontrer," and it is certainly true that almost important discoveries are made in this fashion.

Quoted in James R. Newman
The World of Mathematics
Volume II (p. 919)

Kasner, Edward
Newman, James R.
The mathematician is still regarded as the hermit who knows little of the ways of life outside his cell, who spends his time compounding incredible and incomprehensible theorems in a strange, clipped, unintelligible jargon.

Mathematics and the Imagination (p. xiii)

Keyser, Cassius J.
To think the thinkable—that is the mathematician's aim.

Hibbert Journal
The Universe and Beyond
Volume 3, 1904–1905 (p. 312)

King, Jerry P.
A mathematician, like everyone else, lives in the real world. But the objects with which he works do not. They live in that other place—the mathematical world. Something else lives here also. It is called *truth*.

The Art of Mathematics
Chapter 2 (p. 29)

A mathematician, however, almost always works alone . . . When a mathematician works at mathematics he sits alone in his study staring at equations scribbled on his blackboard or at a dog-eared reprint of the research paper whose results he is trying to extend. It is quiet work, like writing poetry, and includes lots of "dead time" when the mathematician, like the poet, does nothing but sit and stare at the blank page. When you walk in on a research mathematician and find him reclining with his feet up, gazing wistfully out the window, what you say is: "Sorry, I didn't know you were working." Because he probably is.

The Art of Mathematics
Chapter 2 (pp. 36–7)

Kline, Morris
Mathematicians have always constituted a clannish, elitist, snobbish, highly individualistic community in which status is determined, above all, by the presumed importance of original contributions to mathematics; and in which the greatest rewards are bestowed upon those who, at least in the opinion of their peers, will leave a permanent mark on its evolution.

Why the Professor Can't Teach (p. 240)

Langer, Susanne K.
Mathematicians are rarely practical people, or good observers of events. They are apt to be cloistered souls, like philosophers and theologians.

Philosophy in a New Key
Chapter 1 (p. 19)

Lebesgue, Henri
Mathematicians have never been in full agreement on their science, though it is said to be the science of self-evident verities—absolute, indisputable and definitive. They have always been in controversy over the developing aspects of mathematics, and they have always considered their own age to be a period of crisis.

Quoted in Lucienne Felix
The Modern Aspect of Mathematics (p. 3)

In my opinion a mathematician, in so far as he is a mathematician, need not preoccupy himself with philosophy—an opinion, moreover, which has been expressed by many philosophers.

Quoted in E.T. Bell
Men of Mathematics (p. xvii)

Lehmer, Derrick Henry
The real difficulty lies in the fact that only a finite number of angels can dance on the head of a pin, whereas the mathematician is more apt to be interested in the infinite angel problem only.

Bulletin of the American Mathematical Society
Mechanized Mathematics
Volume 72 Number 5, September 1966 (p. 744)

Lewis, Sinclair
"The regularity of the rate at which the strepolysin disappears suggests that an equation may be found—"

"Then why did you not make the equation?"

"Well—I don't know. I wasn't enough of a mathematician."

"Then you should not have published till you knew your math!"

Arrowsmith
Chapter XXVI (p. 288)

Locke, John

As a help to this I think it may be proposed that, for the saving of the long progression of the thoughts to remote and first principles in every case, the mind should provide itself several stages, that is to say, intermediate principles, which it might have recourse to in the examining of those positions that come in its way. These, though they are not self-evident principles, yet, if they have been made out from them by a wary and unquestionable deduction, may be depended on as certain and infallible truths and serve as unquestionable truths to prove other points depending on them by nearer and shorter views than remote and general maxims. These may serve as landmarks to show what lies in the direct way of truth or is quite besides it. And thus mathematicians do, who do not in every new problem run it back to the first axioms through all the whole train of intermediate propositions.

Of the Conduct of the Understanding
Section 21

This is plain and direct sophistry; but I am far from thinking that, wherever it is found, it is made use of with design to deceive and mislead the readers. It is visible that men's prejudices and inclinations by this way impose often upon themselves; and their affection for truth, under their prepossession in favor of one side, is the very thing that leads them from it. Inclination suggests and slides into their discourse favorable terms which introduce favorable ideas, till at last by this means that is concluded clear and evident, thus dressed up, which taken in its native state, by making use of none but the precise determined ideas, would find no admittance at all. The putting these glosses on what they affirm, these (as they are thought) handsome, easy and graceful explications of what they are discoursing on, is so much the character of what is called and esteemed writing well, that it is very hard to think that authors will ever be persuaded to leave what serves so well to propagate their opinions and procure themselves credit in the world for a more jejune and dry way of writing, by keeping to the same terms precisely annexed to the same ideas, a sour and blunt stiffness tolerable in mathematicians only, who force their way and make truth prevail by irresistible demonstration.

Of the Conduct of the Understanding
Section 42

Locke, W.J.

Now all the world understands the irresistible force that compels the poet, at last, to give form to long haunting dreams; the need, also, of the astronomer to crystallize the results of his discoveries and formulate his epoch-making theories; but the passion of the mathematician to do the same is not so easily comprehensible. For years Baltazar had dreamed of an exhaustive and monumental treatise on the Theory of

Groups which would revolutionize the study of the higher mathematics, a gorgeous vision, the mere statement of which must leave the ordinary being cold and the first attempt at explanation petrify him with its icy unintelligibility.

The House of Baltazar
Chapter IV (p. 47)

Mencken, H.L.

Even if we admit that mathematicians are of great value to the world, the fact remains that there are many charlatans (*circulatores*) among them. They talk too much of their discoveries, and nothing grieves them more than to see some other mathematician get ahead of them. How vast is their joy when they solve a problem within the time limits set by him who posited it! They use up every ounce of their energy in that struggle for fame.

Quoted in Johann B. Mencke
The Charlatanry of the Learned
Lecture II, (p. 152, fn 74)

Mordell, Louis Joel

No one will get very far or become a real mathematician without certain indispensable qualities. He must have hope, faith, and curiosity, and prime necessity is curiosity.

Reflections of a Mathematician (p. 7)

Morley, Christopher

Sweep the pale hair, like wings, above the ears;
    Whittle the nose, and carve and bone the jaw;
Blank the studying eyes, till human fears
    Eliminate in universal law.
Slack the mortal shirt, stiffen the hands,
    Holding the dear old pipe, half-smoked, unlit—
So, lovingly, we loose Orion's Bands
    And write equation with the Infinite.

The Ballad of New York, New York and Other Poems 1930–1956
Portrait of a Mathematician

Newton, Sir Isaac

. . . if instead of sending the Observations of seamen to able Mathematicians at Land, the Land would send able Mathematicians to Sea, it would signify much more to the improvement of Navigation and safety of Men's lives and estates on the element.

Quoted in E.G.R. Taylor
The Mathematical Practitioners of Tudor & Stuart England (p. 119)

Oearson, Karl

The mathematician, carried along on his flood of symbols, dealing apparently with purely formal truths, may still reach results of endless importance for our description of the physical universe.

Quoted in John N. Shive and Robert L. Weber
Similarities in Physics (p. 58)

Pascal, Blaise

Mathematicians who are only mathematicians have exact minds, provided all things are explained to them by means of definitions and axioms; otherwise they are inaccurate and insufferable, for they are only right when the principles are quite clear.

Pensées
Section I, 1

Plato

I have hardly ever known a mathematician who was capable of reasoning.

The Republic
Book VII, Section 528

Poe, Edgar Allan

The word "Verse" is used here as the term most convenient for expressing, and without pedantry, all that is involved in the consideration of rhythm, rhyme, meter, and versification . . . the subject is exceedingly simple; one tenth of it, possibly, may be called ethical; nine tenths, however, appertains to the mathematics.

Quoted in Stanley Gudder
A Mathematical Journey (p. 55)

Poincaré, Henri

Mathematicians do not deal in objects, but in relations between objects; thus, they are free to replace some objects by others so long as the relations remain unchanged. Content to them is irrelevant: they are interested in form only.

Quoted in Tobias Dantzig
Number: The Language of Science (p. 317)

The mathematician does not study pure mathematics because it is useful; he studies it because he delights in it and he delights in it because it is beautiful.

Quoted in H.E. Huntley
The Divine Proportion: a Study in Mathematical Beauty (p. 1)

A scientist worthy of the name, above all a mathematician, experiences in his work the same impressions as an artist; his pleasure is as great and of the same nature.

Quoted in Stanley Gudder
A Mathematical Journey (p. 55)

Pringsheim, Alfred

The true mathematician is always a great deal of an artist, an architect, yes, of a poet. Beyond the real world, though perceptibly connected with it, mathematicians have created an ideal world which they attempt to develop into the most perfect of all worlds, and which is being explored in every direction. None has the faintest conception of this world except him who knows it; only presumptuous ignorance can assert that the mathematician moves in a narrow circle. The truth which he seeks is, to be sure, broadly considered, neither more nor less than consistency; but does not his mastership show, indeed, in this very limitation? To solve questions of this kind he passes unenviously over others.

Jahresberichte der Deutschen Mathematiker Vereinigung
Volume 13, 1904 (p. 381)

Rota, Gian-Carlo

A mathematician's work is mostly a tangle of guesswork, analogy, wishful thinking and frustration, and proof, far from being the core of discovery, is more often than not a way of making sure that our minds are not playing tricks.

Quoted in Philip J. Davis and Reuben Hersh
The Mathematical Experience
Introduction (p.xviii)

Sarton, George

The main source of mathematical invention seems to be within man rather than outside of him: his own inveterate and insatiable curiosity, his constant itching for intellectual adventure; and likewise the main obstacles to mathematical progress seem to be also within himself; his scandalous inertia and laziness, his fear of adventure, his need of conformity to old standards, and his obsession by mathematical ghosts.

The Study of the History of Mathematics (p. 16)

Mathematicians and other scientists, however great they may be, do not know the future. Their genius may enable them to project their purpose ahead of them; it is as if they had a special lamp, unavailable to lesser men, illuminating their path; but even in the most favorable cases the lamp sends only a very small cone of light into the infinite darkness.

The Study of the History of Mathematics (pp. 17–18)

The concatenations of mathematical ideas are not divorced from life, far from it, but they are less influenced than other scientific ideas by accidents, and it is perhaps more possible, and more permissible, for a mathematician than for any other man to secrete himself in a tower of ivory.

The Study of the History of Mathematics (pp. 19–20)

Shulman, Milton
I knew a mathematician who said 'I do not know as much as God. But I know as much as God knew at my age.'

Stop the Week, BBC Radio 4

Siegel, Eli
Soup, used rightly by a mathematician, helps him do better.

Damned Welcome
Aesthetic Realism Maxims
Part II, #373 (p. 148)

Solzhenitsyn, Aleksandr
All my life I have thought of mathematicians as Rosicrucians of some kind, and I always regretted that I never had the opportunity of being initiated into their secrets.

The First Circle
Chapter 9 (p. 39)

Stewart, Dugald
. . . I have never met with a *mere mathematician* who was not credulous to a fault . . .

The Collected Works of Dugald Stewart
Volume IV (p. 209)

Swann, W.F.G.
It has been said that the pure mathematician is never as happy as when he does not know what he is talking about . . .

The Architecture of the Universe (p. 117)

Swift, Jonathan
. . . what I chiefly admired, and thought altogether unaccountable, was the strong disposition I observed in them [the mathematicians of Laputa] towards news and politicks; perpetually enquiring into publick affairs; giving their judgments in matters of state; and passionately disputing every inch of party opinions. I have indeed observed the same disposition among most of the mathematicians I have known in Europe; although I could never discover the least analogy between the two sciences . . .

Gulliver's Travels
Part III, Chapter II

Sylvester, James Joseph

I know, indeed, and can conceive of no pursuit so antagonistic to the cultivation of the oratorical faculty . . . as the study of Mathematics. An eloquent mathematician must, from the nature of things, ever remain as rare a phenomenon as a talking fish, and it is certain that the more anyone gives himself up to the study of oratorical effect the less will he find himself in a fit state to mathematicize. It is the constant aim of the mathematician to reduce all his expressions to their lowest terms, to retrench every superfluous word and phrase, and to condense the Maximum of meaning into the Minimum of language. He has to turn his eye ever inwards, to see everything in its dryest light, to train and inure himself to a habit of internal and impersonal reflection and elaboration of abstract thought, which makes it most difficult for him to touch or enlarge upon of those themes which appeal to the emotional nature of his fellow-men. When called upon to speak in public he feels as a man might do who has passed all his life in peering through a microscope, and is suddenly called upon to take charge of an astronomical observatory. He has to get out of himself, as it were, and change the habitual focus of his vision.

Collected Mathematical Works
Volume III
Address on Commemoration Day at Johns Hopkins University (pp. 72–3)

The mathematician lives long and lives young; the wings of his soul do not early drop off, nor do his pores become clogged with the earthy particles blown from the dusty highways of vulgar life.

Source unknown

Synge, John L.

Mathematicians are human beings.

The Scripta Mathematical Studies Number 2
The Life and Early Works of Sir William Rowan Hamilton (p. 13)

The modern mathematician weaves an intricate pattern of microscopic precision. To him, a false statement—an exception to a general statement—is an unforgiven sin. The heroic mathematician, on the other hand, painted with broad splashes of color, with a grand contempt for singular cases until they could no longer be avoided.

The Scripta Mathematical Studies Number 2
The Life and Early Works of Sir William Rowan Hamilton (p. 16)

Thom, René
Everything considered, mathematicians should have the courage of their most profound convictions and thus affirm that mathematical forms indeed have an existence that is independent of the mind considering them . . . Yet, at any given moment, mathematicians have only an incomplete and fragmentary view of this world of ideas.

American Scientist
Modern Mathematics: An Educational and Philosophical Error?
Volume 59, 1971 (p. 695)

Thompson, Silvanus P.
Once when lecturing to a class he [Lord Kelvin] used the word "mathematician," and then interrupting himself asked his class "Do you know what a mathematician is?" Stepping to the blackboard he wrote upon it:—

$$\int_{-\infty}^{\infty} e^{x^2} \, dx = \sqrt{\pi}.$$

Then putting his finger on what he had written, he turned to his class and said: "A mathematician is one to whom *that* is as obvious as that twice two makes four is to you."

In S.P. Thompson
The Life of William Thomson Baron Kelvin of Largs
Volume II
Views and Opinions (p. 1139)

Thomson, William [Lord Kelvin]
Let us then hear no more nonsense about the interference of mathematicians in matters with which they have no concern; rather let them be lauded for condescending from their proud preeminence to help out of a rut the too ponderous wagon of some scientific brother.

In Joe D. Burchfield
Lord Kelvin and the Age of the Earth (p. 93)

Tomlinson, Henry Major
We may doubt the warranty of the priest, but never that of the mathematician.

All Our Yesterdays
Part I, Chapter Two (p. 10)

Truesdell, Clifford A.
Now a mathematician has a matchless advantage over general scientists, historians, politicians, and exponents of other professions: He can be wrong. *A fortiori*, he can also be right. [. . .] A mistake made by a mathematician, even a great one, is not a "difference of a point of view" or "another interpretation of the data" or a "dictate of a conflicting

ideology", it is a mistake. The greatest of all mathematicians, those who have discovered the greatest quantities of mathematical truths, are also those who have published the greatest numbers of lacunary proofs, insufficiently qualified assertions, and flat mistakes. By attempting to make natural philosophy into a part of mathematics, Newton relinquished the diplomatic immunity granted to non-mathematical philosophers, chemists, psychologists, etc., and entered into the area where an error is an error even if it is Newton's error; in fact, all the more so because it is Newton's error.

The mistakes made by a great mathematician are of two kinds: first, trivial slips that anyone can correct, and, second, titanic failures reflecting the scale of the struggle which the great mathematician waged. Failures of this latter kind are often as important as successes, for they give rise to major discoveries by other mathematicians. One error of a great mathematician has often done more for science than a hundred impeccable little theorems proved by lesser men. Since Newton was as great a mathematician as ever lived, but still a mathematician, we may approach his work with the level, tactless criticism which mathematics demands.

Essays in the History of Mechanics
Reactions of Late Baroque Mechanics to Success, Conjecture,
Error, and Failure in Newton's Principia (p. 140)

Ulam, Stanislaw
Mathematicians, at the outset of their creative work, are often confronted by two conflicting motivations: the first is to contribute to the edifice of existing work—it is there that one can be sure of gaining recognition quickly by solving outstanding problems—the second is the desire to blaze new trails and to create new syntheses. This latter course is a more risky undertaking, the final judgment of value or success appearing only in the future.

Bulletin of the American Mathematical Society
John von Neumann, 1903–1957
Volume 64, Number 3, part 2 (p. 8)

Unknown
The mathematician is either a beggar, a *dunce,* or a visionary, or the three in one.

Quoted in
Edinburgh Review
Volume 52
January 1836 (p. 229)

A binary mathematician
Had the curious erotic ambition
   To know what to do
   With the powers of two
When the two are in proper position.

Quoted in E.O. Parrott (Editor)
The Penguin Book of Limericks
Theory and Practice

A mathematician named Hall
Has a hexahedronical ball,
   And the cube of its weight
   Times his pecker, plus eight,
Is his phone number—give him a call.

Quoted in William S. Baring-Gould
The Lure of the Limerick (p. 166)

You have only to show that a thing is impossible and some mathematician will go and do it.

Quoted in W.W. Sawyer
Prelude to Mathematics (p. 167)

Old mathematicians never die; they just lose some of their functions.

Source unknown

A mathematician is a device for turning coffee into theorems.

Source unknown

A mathematician, a physicist, and an engineer were all given a red rubber ball and told to find the volume. The mathematician carefully measured the diameter and evaluated a triple integral. The physicist filled a beaker with water, put the ball in the water, and measured the total displacement. The engineer looked up the model and serial numbers in his red-rubber-ball table.

Source unknown

A businessman needed to employ a quantitative type person. He wasn't sure if he should get a mathematician, an engineer, or an applied mathematician. As it happened, all the applicants were male. The businessman devised a test. The mathematician came first. Miss How, the administrative assistant took him into the hall. At the end of the hall, lounging on a couch, was a beautiful woman. Miss How said, "You may only go half the distance at a time. When you reach the end, you may kiss our model."

The mathematician explained how he would never get there in a finite number of iterations and politely excused himself.

Then came the engineer. He quickly bounded halfway down the hall, then halfway again, and so on. Soon he declared he was well within accepted error tolerance and grabbed the beautiful woman and kissed her.

Finally it was the applied mathematician's turn. Miss How explained the rules. The applied mathematician listened politely, then grabbed Miss How and gave her a big smooch.

"What was that about?" she cried.

"Well, you see I'm an applied mathematician. If I can't solve the problem, I change it!"

Source unknown

An engineer, physicist, and mathematician are all challenged with a problem: to fry an egg when there is a fire in the house. The engineer just grabs a huge bucket of water, runs over to the fire, and puts it out. The physicist thinks for a long while, and then measures a precise amount of water into a container. He takes it over to the fire, pours it on, and with the last drop the fire goes out. The mathematician pores over pencil and paper. After a few minutes he goes "Aha! A solution exists!" and goes back to frying the egg.

Source unknown

Three men are in a hot-air balloon. Soon, they find themselves lost in a canyon somewhere. One of the three men says, "I've got an idea. We can call for help in this canyon and the echo will carry our voices far."

So he leans over the basket and yells out, "Hellllooooooo! Where are we?" (They hear the echo several times).

Fifteen minutes later, they hear this echoing voice: "Hellllooooooo! You're lost!!"

One of the men says, "That must have been a mathematician."

Puzzled, one of the other men asks, "Why do you say that?"

The reply: "For three reasons: (1) he took a long time to answer, (2) he was absolutely correct, and (3) his answer was absolutely useless."

Source unknown

A physicist, an engineer and a mathematician are all standing around the university flagpole when an English professor stops and asks what they're doing. "Well," says the physicist, "we want to know the height of the flagpole and are discussing formulas we might use to calculate it."

"Watch," says the English professor as he takes down the flagpole. Borrowing a measure tape, he measures and says, "Twenty-four feet, exactly."

Then he re-erects the flagpole and goes on his way.

"Humpphhh!", snorts the mathematician. "Isn't that just like an English professor? We ask him for the height and he gives us the length!"

Source unknown

A mathematician, a biologist and a physicist are sitting in a street cafe watching people going in and coming out of the house on the other side of the street.

First they see two people going into the house. Time passes. After a while they notice three persons coming out of the house.

The physicist: "The measurement wasn't accurate."

The biologist's conclusion: "They have reproduced."

The mathematician: "If now exactly 1 person enters the house then it will be empty again."

Source unknown

An engineer, a physicist, and a mathematician are shown a pasture with a herd of sheep, and told to put them inside the smallest possible amount of fence. The engineer is first. He herds the sheep into a circle and then puts the fence around them, declaring, "A circle will use the least fence for a given area, so this is the best solution."

The physicist is next. She creates a circular fence of infinite radius around the sheep, and then draws the fence tight around the herd, declaring, "This will give the smallest circular fence around the herd."

The mathematician is last. After giving the problem a little thought, he puts a small fence around himself and then declares, "I define myself to be on the outside!"

Source unknown

A doctor, a lawyer and a mathematician were discussing the relative merits of having a wife or a mistress. The lawyer says: "For sure a mistress is better. If you have a wife and want a divorce, it causes all sorts of legal problems."

The doctor says: "It's better to have a wife because the sense of security lowers your stress and is good for your health."

The mathematician says: "You're both wrong. It's best to have both so that when the wife thinks you're with the mistress and the mistress thinks you're with your wife—you can do some mathematics."

Source unknown

Percy P was a mathematician
  whose "pureness" was never denied.
But he found one day, to his sorrow,
  that his theorems had been applied!
He had used all the standard precautions;
  his papers were pointedly dry!
But his own esoteric notation
  had been solved by a physicist spy!

The colloquium buzzed with the gossip;
  he could offer no valid excuse.
Percy P was a traitor of traitors,
  for his work was of PRACTICAL USE!
Nobody dared to defend him.
  Could it be that he'd plead the crime
That his work was just then needed
  to effect quantization of time?

Ignored when he joined conversations;
  one would think that he poisoned the air.
And he felt on his way to the office—
  a new man might be in his chair.
A committee was in operation,
  working twenty four hours a day,
Deleting his name from the journals,
  and throwing his reprints away.

He knew where his future was leading,
  no sense in prolonging the pain;
He left with a handful of papers,
  and never was heard from again.
So take heed all you mathematicians
  who pretend your endeavor is pure;
Tho' your luck may hold for a decade,
  in the end you can never be sure.

POOR PURE PERCY P.
Credited to the *Princeton Tiger*

... what the mathematician predicts today has a habit of becoming what the physicist finds tomorrow.

The Times
October 4, 1934

The polite mathematician says, "You go to infinity—and don't hurry back!"

Source unknown

A MATHEMATICIAN is one who endeavors to secure the greatest possible consistency in his thoughts ad statements by guiding the process of his reasoning into those well-worn tracks by which we pass from one relation among quantities to an equivalent relation. He who has kept his mind always in those paths which have never led him or anyone else to an inconsistent result, and has traversed them so often that the act of passage has become rather automatic than voluntary, is, and knows himself to be, an accomplished mathematician.

Nature
Quaternions
Volume ix, Number 217, December 25, 1873 (p. 137)

The difference between the Pope and a mathematician is that although both make infallible dictums only the mathematician can be checked.

Source unknown

Hey baby, How would you like to join me in some math? We'll add you and me, subtract our clothes, divide your legs, and multiply! Of course, we'll be entirely discrete.

Mathematician's Pick Up Line
Source unknown

Engineers think that equations approximate the real world.

Physicists think that the real world approximates equations.

Mathematicians are unable to make the connection.

Source unknown

A promising Ph.D. candidate was presenting his thesis at his final examination. He proceeded with a derivation and ended up with something like:

$$F = -MA.$$

He was embarrassed, his supervising professor was embarrassed, and the rest of the committee was embarrassed. The student coughed nervously and said "I seem to have made a slight error back there somewhere."

One of the mathematicians on the committee replied dryly, "Either that or an odd number of them!"

Quoted on the Internet

Mathematicians may flatter themselves that they possess new ideas which mere human language is as yet unable to express. Let them make the effort to express these ideas in appropriate words without the aid of symbols, and if they succeed, they will not only lay us laymen under a lasting obligation, but, we venture to say, they will find themselves very much enlightened during the process, and will even be doubtful whether the ideas as expressed in symbols had ever quite found their way out of the equations into their minds.

Book review of Sir W. Thomson and P.G. Tait
Elements of Natural Philosophy
Nature
Volume vii, Number 178, March 27, 1873 (p. 400)

Veblen, Oswald

... let me remind any non-mathematicians ... that when a mathematician lays down the elaborate tools by which he achieves precision in his own domain, he is unprepared and awkward in handling the ordinary tools of language. This is why mathematicians always disappoint the expectation that they will be precise and reasonable and clear-cut in their statements about everyday affairs, and why they are, in fact, more fallible than ordinary mortals.

Science
Geometry and Physics
February 2, 1923

The branch of physics which is called Elementary Geometry was long ago delivered into the hands of mathematicians for the purposes of instruction. But, while mathematicians are often quite competent in their knowledge of the abstract nature of the subject, they are rarely so in their grasp of its physical meaning.

Science
Geometry and Physics
February 2, 1923

The conclusion seems inescapable: that formal logic has to be taken over by the mathematicians. The fact is that there does not exist an adequate logic at the present time, and unless the mathematicians create one, no one else is likely to do so.

Quoted in
A Century of Mathematics in America
Volume II
Retiring address to the AMS 1924 (p. 219)

von Hardenberg, Friedrich

The mathematicians are the only happy ones. He who does not reach for a mathematical book with devotion and read it as the word of God does not understand it.

Quoted in Walter R. Fuchs
Mathematics for the Modern Mind (p. 18)

Walther, Hans
Take the mathematician away: He is a stupid augur, blind prophet, a crazy soothsayer. Man may know the present; only God can foresee the future.

In Jan Gullberg
Mathematics from the Birth of Numbers (p. 17)

Weierstrass, Karl
A mathematician who is not also something of a poet will never be a complete mathematician.

Quoted in Oswald Spengler
The Decline of the West
Volume I
Chapter II, section iv (p. 62)

Weil, André
Rigor is to the mathematician what morality is to man. It does not consist in proving everything, but in maintaining a sharp distinction between what is assumed and what is proved, and in endeavoring to assume as little as possible at every stage.

American Mathematical Monthly
Mathematical Teaching in Universities
Volume 61, Number 1, January 1954 (p. 35)

But, if logic is the hygiene of the mathematician, it is not his source of food; the great problems furnish the daily bread on which he thrives.

American Mathematical Monthly
The Future of Mathematics
Volume 57, Number 5, May 1950 (p. 297)

Weiner, Norbert
One of the chief duties of the mathematician in acting as an advisor to scientists . . . is to discourage them from expecting too much from mathematics.

Quoted in Douglas M. Campbell and John C. Higgins
Mathematics: People, Problems, Results
Volume III (p. 29)

Weyl, Hermann
It cannot be denied, however, that in advancing to higher and more general theories the inapplicability of the simple laws of classical logic eventually results in an almost unbearable awkwardness. And the mathematician watches with pain the larger part of his towering edifice which he believed to be built of concrete dissolve into mist before his eyes.

Philosophy of Mathematics and Natural Science
Intuitive Mathematics (p. 54)

White, William Frank
Behind the artisan is the chemist, behind the chemist a physicist, behind the physicist a mathematician.

Scrap-book of Elementary Mathematics (p. 217)

Wiener, Norbert
We mathematicians who operate with nothing more expensive than paper and possibly printer's ink are quite reconciled to the fact that, if we are working in a very active field, our discoveries will commence to be obsolete at the moment they are written down or even at the moment they are conceived. We know that for a long time everything we do will be nothing more than the jumping off point for those who have the advantage of already being aware of our ultimate results. This is the meaning of the famous apothegm of Newton, when he said, "If I have seen further than other men, it is because I have stood on the shoulders of giants".

I Am a Mathematician (p. 266)

Wittgenstein, Ludwig
The mathematician is an inventor, not a discoverer.

Remarks on the Foundations of Mathematics
Appendix I, 167 (p. 47e)

Mathematicians practice absolute freedom.
Henry Adams – (See p. 149)

MATHEMATICS

Abel, Niels Henrik
It appears to me that if one wants to make progress in mathematics, one should study the masters and not the pupils.

Quoted in Ø. Ore
Niels Henrik Abel, Mathematician Extraordinary (p. 138)

Adams, Henry
In the one branch he most needed—mathematics—barring the few first scholars, failure was so nearly universal that no attempt at grading could have had value, and whether he stood fortieth or ninetieth must have been an accident or the personal favor of the professor. Here his education failed lamentably. At best he could never have been a mathematician; at worst he would never have cared to be one; but he needed to read mathematics, like any other universal language, and he never reached the alphabet.

The Education of Henry Adams
Chapter IV (p. 60)

Allen, Woody
Standard mathematics has recently been rendered obsolete by the discovery that for years we have been writing the numeral five backward. This has led to a reevaluation of counting as a method of getting from one to ten.

Getting Even (p. 58)

Anglin, W.S.
Mathematics is not a careful march down a well-cleared highway, but a journey into a strange wilderness, where the explorers often get lost. Rigour should be a signal to the historian that the maps have been made, and the real explorers have gone elsewhere.

Mathematical Intelligencer
Mathematics and History
Volume 14, Number 4 (p. 10)

Apostle, H.G.

That mathematics is a theoretical science is evident from those who have pursued that science; they have been lovers of wisdom and have sought to discover eternal truths . . . those who have investigated the objects of mathematics have done so not for gain or use but for the sake of truth.

Aristotle's Philosophy of Mathematics
Chapter I (p. 3)

Aristo, Chian

Those who occupy themselves with Mathematics to the neglect of Philosophy, are like the wooers of Penelope, who, unable to attain the mistress, contend themselves with the maids.

Quoted in
Edinburgh Review
Volume 52
January 1836 (p. 229)

Aristotle

There are things which seem incredible to most men who have not studied mathematics.

Quoted in Stanley Gudder
A Mathematical Journey (p. 358)

Ascham, Roger

Mark all Mathematical heads [he continued] which be wholly and only bent on these sciences, how solitary they be themselves, how unfit to live with others, how unapt to serve the world.

Quoted in E.G.R. Taylor
The Mathematical Practitioners of Tudor and Stuart England
Introduction (p. 5)

Asimov, Isaac

I'm saying *suppose*. In mathematics, we say 'suppose' all the time and see if we can end up with something patently untrue or self-contradictory
. . .

Prelude to Foundation (p. 206)

Bacon, Francis

For it being the nature of the mind of man (to the extreme prejudice of knowledge) to delight in the spacious liberty of generalities, as in a champain region, and not in the inclosures of particularity, the mathematics of all other knowledge were the goodliest fields to satisfy that appetite.

The Advancement of Learning
Book II, Chapter VIII, section 1

For many parts of nature can neither be invented with sufficient subtlety, nor demonstrated with sufficient perspicuity, nor accommodated unto use with sufficient dexterity, without the aid and intervening of mathematics . . .

The Advancement of Learning
Book II, Chapter VIII, section 2

In mathematics I can report no deficience, except it be that men do not sufficiently understand the excellent use of the pure mathematics.

The Advancement of Learning
Book II, Chapter VIII, section 2

As Physic advances farther and farther every day and develops new axioms, it will require fresh assistance from Mathematic.

The Advancement of Learning
Book III, Chapter VI

So if a man's wit be wandering, let him study the mathematics; for in demonstrations, if his wit be called away never so little, he must begin again.

Bacon's Essays
Of Studies

Bacon, Roger
There are four great sciences . . . Of these sciences the gate and key is mathematics, which the saints discovered at the beginning of the world . . .

Opus Majus
Part 4, Chapter 1 (p. 116)

. . . mathematics is absolutely necessary and useful to other sciences.

Opus Majus
Part 4, Chapter 3 (p. 126)

Bain, Alexander
Those that can readily master the difficulties of Mathematics find a considerable charm in the study, sometimes amounting to fascination. This is far from universal; but the subject contains elements of strong interest of a kind that constitutes the pleasures of knowledge. The marvelous devices for solving problems elate the mind with the feeling of intellectual power; and the innumerable constructions of the science leave us lost in wonder.

Education as a Science (p. 153)

Barnett, P.A.
The strength of mathematics is derived from the very fact that its truths are detachable by abstraction from the concrets in which they exist for our senses. We argue *in vacuo*, so to speak, without any possibility of error arising from the accidents of individual experience, prejudice, opinion, or the imperfection of our senses.

Common Sense in Education and Teaching (p. 222)

Barrow, Isaac
Mathematics—the unshaken Foundation of Sciences, and the plentiful Fountain of Advantage to human affairs.

Quoted in Carl B. Boyer
A History of Mathematics (p. 404)

Bartlett, Albert A.
The greatest shortcoming of the human race is man's inability to understand the exponential function.

The Physics Teacher
The Exponential Function
Volume 14, Number 7, October 1976 (p. 394)

Bateson, Gregory
Still more astonishing is that world of rigorous fantasy we call mathematics.

Steps to an Ecology of the Mind
Redundancy and Coding (p. 422)

Begley, Sharon
Pure mathematics is a sucker's game. It lures the curious and the confident with its seeming simplicity only to make them look like fools.

Newsweek
New Answer for an Old Question
July 5, 1993 (p. 52)

Bell, Eric T.
For my own part I have swallowed mathematics enough in my life to be immune to just one more dose, and I shall continue to get smallpox vaccinations whenever I contemplate a vacation in any of the filthier parts of the North American continent.

Mathematics: Queen and Servant of Science
Choice and Chance (p. 381)

'Obvious' is the most dangerous word in mathematics.

Mathematics: Queen and Servant of Science
Points of View (p. 16)

'If it is not abstract it is not mathematics' might be taken as a touchstone for discriminating between mathematics and other departments of precise investigation.

> *Mathematics: Queen and Servant of Science*
> Abstraction and Prediction (p. 259)

Mathematics by itself has seldom got very far in the exploration of nature, as is attested by the numerous attempts of pure mathematicians of the past and present to resolve the universe with pencil and paper.

> *Mathematics: Queen and Servant of Science*
> A Metrical Universe (p. 211)

The task of cleaning up mathematics and salvaging whatever can be saved from the wreckage of the past twenty years will probably be enough to occupy one generation.

> *Debunking Science*

Mathematics has a light and wisdom of its own, above any possible applications to science, and it will richly reward any intelligent human being to catch a glimpse of what mathematics means to itself. This is not the old doctrine of art for art's sake; it is art for humanity's sake.

> *Men of Mathematics* (p. 4)

The longer mathematics lives the more abstract—and therefore, possibly also the more practical—it becomes.

> *The Mathematical Intelligencer*
> Volume 13, Number 1, Winter 1991

When we begin unraveling a particular thread in the history of mathematics we soon get a discouraged feeling that mathematics itself is like a vast necropolis to which constant additions are being made for the eternal preservation of the newly dead. The recent arrivals, like some of the few who were shelved for perpetual remembrance 5000 years ago, must be so displayed that they shall seem to retain the full vigor of the manhood in which they died; in fact the illusion must be created that they have not yet ceased living. And the deception must be so natural that even the most skeptical archaeologist prowling through the mausoleums shall be moved to exclaim with living mathematicians themselves that mathematical truths are immortal, imperishable; the same yesterday, today, and forever, the very stuff of which eternal verities are fashioned and the one glimpse of changelessness behind all the recurrent cycles of birth, death, and decay our race has ever caught.

> *Men of Mathematics* (p. 11)

Even stranger things have happened; and perhaps the strangest of all is the marvel that mathematics should be possible to a race akin to the apes.

The Development of Mathematics
Uncertainties and Probabilities (p. 546)

Bellman, Richard
Mathematics makes natural questions precise.

Eye of the Hurricane: An Autobiography (p. 114)

Bentley, Arthur
The every-day language reeks with philosophies . . . It shatters at every touch of advancing knowledge. At its heart lies paradox.

The language of mathematics, on the contrary, stands and grows in firmness. It gives service to men beyond all other language.

Linguistic Analysis of Mathematics
Foreword (p. viii)

Bergson, Henri
. . . calculation touches, at most, certain phenomena of organic *destruction*. Organic *creation*, on the contrary, the evolutionary phenomena which properly constitute life, we cannot in any way subject to mathematical treatment.

Creative Evolution (p. 20)

Bers, Lipman
. . . mathematics is very much like poetry . . . what makes a good poem— a great poem—is that there is a large amount of thought expressed in very few words. In this sense formulas like

$$e^{\pi i} + 1 = 0$$

or

$$\int_{-\infty}^{+\infty} e^{-x^2} \, dx = \sqrt{\pi}$$

are poems.

Quoted in D. Albers, G. Alexanderson, and C. Reid (Editors)
More Mathematical People
Lipman Bers (p. 16)

Bishop, Errett
Mathematics belongs to man, not to God. We are not interested in properties of the positive integers that have no descriptive meaning for finite man. When a man proves a positive integer to exist, he should

show how to find it. If God has mathematics of his own that need to be done, let him do it himself.

Foundations of Constructive Analysis
Chapter 1 (p. 2)

Bôcher, Maxime

I like to look at mathematics almost more as an art than as a science; for the activity of the mathematician, constantly creating as he is, guided although not controlled by the external world of senses, bears a resemblance, not fanciful, I believe, but real, to the activities of the artist, of a painter, let us say. Rigorous deductive reasoning on the part of the mathematician may be likened here to the technical skill in drawing on the part of the painter. Just as one cannot become a painter without a certain amount of skill, so no one can become a mathematician without the power to reason accurately up to a certain point. Yet these qualities, fundamental though they are, do not make a painter or a mathematician worthy of the name, nor indeed are they the most important factors in the case. Other qualities of a far more subtle sort, chief among which in both cases is imagination, go to the making of a good artist or a good mathematician.

Bulletin of the American Mathematical Society
2nd Series
The Fundamental Conceptions and Methods in Mathematics
Volume 11, 1904 (p. 133)

Bochner, Salomon

Mathematics is a form of poetry which transcends poetry in that it proclaims a truth; a form of reasoning which transcends reasoning in that it wants to bring about the truth it proclaims; a form of action, of ritual behavior, which does not find fulfillment in the act but must proclaim and elaborate a poetic form of truth.

The Role of Mathematics in the Rise of Science (p. 14)

The word "mathematics" is a Greek word, and, by origin, it means "something that has been learned or understood," or perhaps "acquired knowledge," and perhaps even, somewhat against grammar, "acquirable knowledge," that is, "learnable knowledge," that is, "knowledge acquirable by learning."

The Role of Mathematics in the Rise of Science (pp. 24–5)

Bowditch, N.

I never came across one of Laplace's '*Thus it plainly appears*' without feeling sure that I have hours of hard work before me to fill up the chasm and find out and show *how* it plainly appears.

Quoted in Florian Cajori
The Teaching and History of Mathematics in the United States (p. 104)

Boyer, Carl
. . . mathematics is an aspect of culture as well as a collection of algorithms.

> *The History of the Calculus and Its Conceptual Development*
> Preface to second printing

Mathematics is neither a description of nature nor an explanation of its operation; it is not concerned with physical motion or with the metaphysical generation of quantities. It is merely the symbolic logic of possible relations, and as such is concerned with neither approximate nor absolute truth, but only with hypothetical truth. That is, mathematics determines what conclusions will follow logically from given premises. The conjunction of mathematics and philosophy, or of mathematics and science is frequently of great service in suggesting new problems and points of view.

> *The History of the Calculus and Its Conceptual Development* (p. 308)

Bragdon, Claude
Mathematics is the handwriting on the human consciousness of the very Spirit of Life itself.

> Quoted in John D. Barrow
> *Pi in the Sky* (p. 21)

Bridges, Robert
. . . and see how Mathematick rideth as a queen
Cheer'd on her royal progress thru'out nature's realm.

> *The Testament of Beauty*
> IV, l. 856–7

Bridgman, Percy William
It is the merest truism, evident at once to unsophisticated observation, that mathematics is a human invention.

> *The Logic of Modern Physics* (p. 60)

As at present constructed, mathematics reminds one of the loquacious and not always coherent orator, who was said to be able to set his mouth going and go off and leave it.

> *The Logic of Modern Physics* (p. 63)

Bronowski, Jacob
. . . I find both a special pleasure and constraint in describing the progress of mathematics, because it has been part of so much speculation: a ladder for mystical as well as rational thought in the intellectual ascent of man.

> *The Ascent of Man* (p. 155)

Browne, Sir Thomas

All things began in order; so shall they end, and so shall they begin again; according to the ordainer of order and the mystical mathematics of the City of Heaven.

The Garden of Cyrus
Chapter V (p. 114)

Buchanan, Scott

Mathematics then becomes the ladder by which we all may climb into the heaven of perfect insight and eternal satisfaction, and the solution of arithmetic and algebraic problems is connected with the salvation of our souls.

Poetry and Mathematics (p. 37)

. . . mathematics and poetry move together between two extremes of mysticism, the mysticism of the common-place where ideas illuminate and create facts, and the mysticism of the extraordinary where God, the Infinite, the Real, poses the riddles of desire and disappointment, sin and salvation, effort and failure, question and paradoxical answer . . .

Poetry and Mathematics (p. 42)

The structures with which mathematics deals are like lace, like the leaves of trees, like the play of light and shadow on a meadow, or on a human face.

Quoted in Nathan A. Court
Mathematics in Fun and in Earnest

Bullock, James

Mathematics is not a way of hanging numbers on things so that quantitative answers to ordinary questions can be obtained. It is a language that allows one to think about extraordinary questions.

American Mathematical Monthly
Literacy in the Language of Mathematics
Volume 101, Number 8, October 1994 (p. 737)

Butler, Nicholas

Modern mathematics, that most astounding of intellectual creations, has projected the mind's eye through infinite time and the mind's hand into boundless space.

The Meaning of Education and Other Essays and Addresses (p. 44)

Butler, Samuel
In mathematics he was greater
Than Tycho Brahe or Erra Pater;
For he, by geometric scale,
Could take the size of pots of ale;
Resolve by sines and tangents straight
If bread or butter wanted weight;
And wisely tell you what hour o' th' day
The clock does strike, by algebra.

The Poetical Works of Samuel Butler
Volume I
Hudibras
Part I, Canto I, l. 119–26

Cajori, Florian
One of the most baneful delusions by which the minds, not only of
students, but even of many teachers of mathematics in our classical
colleges, have been afflicted is, that mathematics can be mastered by
the favored few, but lies beyond the grasp and power of the ordinary
mind.

The Teaching and History of Mathematics in the United States (p. 100)

The history of mathematics is important also as a valuable contribution
to the history of civilization. Human progress is closely identified with
scientific thought. Mathematical and physical researches are a reliable
record of intellectual progress.

History of Mathematics (p. 4)

Cantor, Georg Ferdinand Ludwig Phillip
Je le vois, mais je ne le crois pas!
[I see it, but I don't believe it!]

Letter to R. Dedekind
June 20, 1877
regarding results of "Theory of Manifolds"
Gesammelte Abhandlungen Mathematischen and Philosophischen Inhalts (p. 458)

The essence of mathematics is its freedom.

Mathematische Annalen
Bd. 21 (p. 564)

Carmichael, R.D.
Mathematics and poetry lie, if not on, at least not far from the extremes,
the one of systematic and the other of unsystematic thought, and thus
are about as far removed as possible one from the other.

The Logic of Discovery (p. 244)

Mathematics, by exhibiting a body of truth which can live through millenniums without needed connections, and at the same time can grow in magnitude and range and interest, has given the human spirit new ground for believing in itself and for rejoicing in its power of consistent thought.

The Logic of Discovery (pp. 263–4)

Carus, Paul

There is no science which teaches the harmonies of nature more clearly than mathematics . . .

Quoted in William Symes Andrews
Magic Squares and Cubes
Introduction

Casson, Stanley

The nearer man approaches mathematics the farther away he moves from the animals.

Progress and Catastrophe
Chapter VI (p. 93)

Chapman, C.H.

There is probably no other science which presents such different appearances to one who cultivates it and one who does not, as mathematics. To [the noncultivator] it is ancient, venerable, and complete; a body of dry, irrefutable, unambiguous reasoning. To the mathematician, on the other hand, his science is yet in the purple of bloom of vigorous youth, everywhere stretching out after the "attainable but unattained," and full of the excitement of nascent thoughts; its logic is beset with ambiguities, and its analytic processes, like Bunyan's road, have a quagmire on one side and a deep ditch on the other, and branch off into innumerable by-paths that end in a wilderness.

New York Mathematical Society Bulletin
The Theory of Transformation Groups
Volume 2, Number 14, 1893 (p. 61)

Chesterton, Gilbert Keith

. . . the common idea that mathematics is a dull subject whereas the testimony of all those who have any dealings with it shows that it is one of the most thrilling and tantalising and enchanting subjects in the world.

Lunacy and Letters
A Defence of Bores (pp. 58–9)

Churchill, Winston Spencer

I had a feeling once about Mathematics—that I saw it all. Depth beyond Depth was revealed to me—the Byss and the Abyss. I saw—as one might see the transit of Venus or even the Lord Mayor's Show—a quantity passing through infinity and changing its sign from plus to minus. I saw exactly why it happened and the tergiversation was inevitable—but it was after dinner and I let it go.

My Early Life: A Roving Commission
A Roving Commission (p. 26)

Some of my cousins who had the great advantage of University education used to tease me with arguments to prove that nothing has any existence except what we think of it . . . These amusing mental acrobatics are all right to play with. They are perfectly harmless and perfectly useless . . . I always rested on the following argument . . .

We look up to the sky and see the sun. Our eyes are dazzled and our senses record the fact. So here is this great sun standing apparently on no better foundation than our physical senses. But happily there is a method, apart altogether from our physical senses, of testing the reality of the sun. It is by mathematics. By means of prolonged processes of mathematics, entirely separate from the senses, astronomers are able to calculate when an eclipse will occur. They predict by pure reason that a black spot will pass across the sun on a certain day. You go and look, and your sense of sight immediately tells you that their calculations are vindicated. So here you have the evidence of the senses reinforced by the entirely separate evidence of a vast independent process of mathematical reasoning. We have taken what is called in military map-making "a cross bearing." . . .

When my metaphysical friends tell me that the data on which the astronomers made their calculations, were necessarily obtained originally through the evidence of the senses, I say, "no." They might, in theory at any rate, be obtained by automatic calculating-machines set in motion by the light falling upon them without admixture of the human senses at any stage. When it is persisted that we should have to be told about the calculations and use our ears for that purpose, I reply that the mathematical process has a reality and virtue in itself, and that once discovered it constitutes a new and independent factor. I am also at this point accustomed to reaffirm with emphasis my conviction that the sun is real, and also that it is hot—in fact hot as Hell, and that if the metaphysicians doubt it they should go there and see.

My Early Life: A Roving Commission (pp. 123–4)

Coleridge, Samuel Taylor

I have often been surprised that Mathematics, the quintessence of Truth, should have found admirers so few and so languid.

The Complete Poetical Works of Samuel Taylor Coleridge
Volume I
A Mathematical Problem

Some people have contended that mathematics ought to be taught by making illustrations obvious to the senses. Nothing can be more absurd or injurious: It ought to be our never-ceasing effort to make people think, not feel.

Lectures and Notes on Shakespeare and Other English Poets
Lecture II (p. 52)

Colton, Charles Caleb

The science of mathematics performs more than it *promises*.

Lacon (p. 181)

The study of mathematics, like the Nile, begins in minuteness, but ends in magnificence . . .

Lacon (p. 181)

Comte, Auguste

. . . it is only through Mathematics that we can thoroughly understand what true science is. Here alone can we find in the highest degree simplicity and severity of scientific law, and such abstraction as the human mind can attain. Any scientific education setting forth from any other point, is faulty in its basis.

The Positive Philosophy
Volume I
Book I, Chapter I (p. 39)

In the present state of our knowledge we must regard Mathematics less as a constituent part of natural philosophy than as having been, since the time of Descartes and Newton, the true basis of the whole natural philosophy; though it is, exactly speaking, both the one and the other. To us it is of less value for the knowledge of which it consists, substantial and valuable as that knowledge is, than as being the most powerful instrument that the human mind can employ in the investigation of the laws of natural phenomena.

The Positive Philosophy
Volume I
Introduction, Chapter II (p. 32)

Geometrical and Mechanical phenomena are the most general, the most simple, the most abstract of all,—the most irreducible to others, the most independent of them; serving, in fact, as a basis to all others. It follows that the study of them is an indispensable preliminary to that of all others. Therefore must Mathematics hold the first place in the hierarchy of the sciences, and be the point of departure of all Education, whether general or special.

The Positive Philosophy
Volume I
Introduction, Chapter II (p. 33)

Conant, James B.
Mathematics and measurement are not to be unduly worshipped, nor can they be neglected by even the lay observer.

Science and Common Sense
Chapter Six (p. 163)

Conrad, Joseph
The demonstration must be against learning science. But not every science will do. The attack must have all the shocking senselessness of gratuitous blasphemy. Since bombs are your means of expression, it would be really telling if one could throw a bomb into pure mathematics. But that is impossible . . . What do you think of having a go at astronomy?

The Secret Agent
Chapter II (p. 38)

Copernicus, Nicolaus
. . . if the worth of the arts were measured by the matter with which they deal, this art—which some call astronomy, others astrology, and many of the ancients the consummation of mathematics—would be by far the most outstanding. This art which is as it were the head of all the liberal arts and the one most worthy of a free man leans upon nearly all the other branches of mathematics. Arithmetic, geometry, optics, geodesy, mechanics, and whatever others, all offer themselves in its service.

On the Revolutions of the Heavenly Spheres
Book One
Introductory paragraph one

d'Alembert, Jean le Rond
We shall content ourselves with the remark that if mathematics (as is asserted with sufficient reason) only makes straight the minds which are without bias, so they only dry up and chill the minds already prepared for this operation by nature.

Quoted in
Edinburgh Review
Volume 52
January 1836 (p. 224)

da Vinci, Leonardo
Therefore, O students, study mathematics, and do not build without foundations.

The Notebooks of Leonardo da Vinci
Volume I
Philosophy (p. 88)

He who blames the supreme certainty of mathematics feeds on confusion, and will never impose silence upon the contradictions of the sophistical sciences, which occasion a perpetual clamour.

The Notebooks of Leonardo da Vinci
Volume I
Philosophy (p. 88)

Dantzig, Tobias
Mathematics is the supreme judge; from its decisions there is no appeal.

Number: The Language of Science
Chapter 12, 12 (p. 245)

It is a remarkable fact that the mathematical inventions which have proved to be most accessible to the masses are also those which exercised the greatest influence on the development of pure mathematics.

Number: The Language of Science (p. 192)

Banish the infinite process, and mathematics pure and applied is reduced to the state in which it was known to the pre-Pythagoreans.

Number: The Language of Science (p. 137)

Darwin, Charles
Every *new* body of discovery is mathematical in form, because there is no other guidance we can have.

Quoted in E.T. Bell
Men of Mathematics (p. xvii)

Davies, Robertson
Although I am almost illiterate mathematically, I grasped very early in life that any one who can count to ten can count upward indefinitely if he is fool enough to do so.

The Table Talk of Samuel Marchbanks
Of the Conservation of Youth (pp. 27–8)

Davis, Chandler
Mathematics is armchair science.

> In R.S. Cohen, J.J. Stachel and M.W. Wartofsky
> *Boston Studies in the Philosophy of Science*
> Volume XV
> Materialist Mathematics (p. 38)

Davis, Philip J.
One of the endlessly alluring aspects of mathematics is that its thorniest paradoxes have a way of blooming into beautiful theories.

> *Scientific American*
> Numbers
> Volume 211 Number 3, September 1964 (p. 55)

Davis, Philip J.
Hersh, Reuben
It seems certain that there is a limit to the amount of living mathematics that humanity can sustain at any time. As new mathematical specialties arise, old ones will have to be neglected.

> *The Mathematical Experience*
> How Much Mathematics Can There Be? (p. 25)

de Bruijn, N.G.
Usually in mathematics one has to choose between saying more and more about less and less on one hand, and saying less and less about more and more on the other.

> *Asymptotic Methods in Analysis* (p. v)

de Jouvenel, Bertrand
. . . the social scientist who lacks a mathematical mind and regards a mathematical formula as a magic recipe, rather than as the formulation of a supposition, does not hold forth much promise. A mathematical formula is never more than a precise statement. It must not be made into a Procrustean bed—and that is what one is driven to by the desire to make the future known to us, and those who think it can would once have believed in witchcraft. The chief merit of mathematicization is that it compels us to become conscious of what we are assuming.

> *The Art of Conjecture* (p. 173)

de Morgan, Augustus
There are no limits in mathematics, and those that assert there are, are infinite ruffians, ignorant, lying blackguards. There is no differential calculus, no Taylor's theorem, no calculus of variations, &c. in mathematics. There is no quackery whatever in mathematics; no % equal to anything.

> *A Budget of Paradoxes*
> Volume I (p. 262)

The *pseudomath* is a person who handles mathematics as the monkey handled the razor.

A Budget of Paradoxes
Volume II (p. 338)

The *graphomath* is a person who, having no mathematics, attempts to describe a mathematician.

A Budget of Paradoxes
Volume II (p. 340)

de Stael, Madame
The mathematics lead us to lay out of account all that is not proved; while the primitive truths, those which sentiment and genius apprehend, are not susceptible of demonstration.

Quoted in
Edinburgh Review
Volume 52
January 1836 (p. 248)

Dehn, Max
Mathematics is the only instructional material that can be presented in an entirely undogmatic way.

The Mathematical Intelligencer
The Mentality of the Mathematician
Volume 5, Number 2, 1983 (p. 73)

Descartes, René
But as I considered the matter carefully it gradually came to light that all those matters only were referred to Mathematics in which order and measurement are investigated, and it makes no difference whether it be in numbers, figures, stars, sounds or any other objects that the question of measurement arises.

Rules for the Direction of the Mind
Rule IV

Dieudonné, Jean
On foundations we believe in the reality of mathematics, but of course when philosophers attack us with their paradoxes we rush to hide behind formalism and say, "Mathematics is just a combination of meaningless symbols," and then we bring out Chapters 1 and 2 on set theory. Finally we are left in peace to go back to our mathematics and do it as we have always done, with the feeling each mathematician has that he is working on something real. This sensation is probably an illusion, but is very convenient.

American Mathematical Monthly
The Works of Nicholas Bourbaki
Volume 77, 1970 (p. 134)

Dirac, Paul Adrien Maurice
God used beautiful mathematics in creating the world.

Quoted in Heinz R. Pagels
The Cosmic Code (p. 191)

Our feeble attempts at mathematics enable us to understand a bit of the universe, and as we proceed to develop higher and higher mathematics we can hope to understand the universe better.

Scientific American
The Evaluation of the Physicist's Picture of Nature
Volume 208, Number 5, May 1963 (p. 53)

Dodgson, Charles L.
The bottom line for mathematicians is that the architecture has to be right. In all the mathematics that I did, the essential point was to find the right architecture. It's like building a bridge. Once the main lines of the structure are right, then the details miraculously fit. The problem is the overall design.

The College Mathematics Journal
Freeman Dyson: Mathematician, Physicist, and Writer
Volume 25, Number 1, January 1994

It may well be doubted whether, in all the range of Science, there is any field so fascinating to the explorer—so rich in hidden treasures—so fruitful in delightful surprises—as that of Pure Mathematics.

A New Theory of Parallels
Introduction (p. xv)

Eddington, Sir Arthur Stanley
Proof is an idol before whom the pure mathematician tortures himself.

The Nature of the Physical World (p. 337)

The solution goes on famously; but just as we have got rid of the other unknowns, behold! V disappears as well, and we are left with the indisputable but irritating conclusion—

$$0 = 0$$

This is a favorite device that mathematical equations resort to, when we propound stupid questions.

The Nature of the Physical World (p. 30)

The mathematics is not there till we put it there.

The Philosophy of Physical Science
Chapter IX (p. 136)

The Editor of The Monist
There is no prophet that preaches the superpersonal God more plainly than mathematics.

The Monist
The God Problem
Volume XVI, Number 1, January 1906 (p. 147)

Edwards, Harold M.
Mathematics, like philosophy, is virtually inseparable from its history.

Quoted in Lynn Arthur Steen
Mathematics Tomorrow (p. 108)

Edwards, Tyron
The study of mathematics is like climbing up a steep and craggy mountain; when once you reach the top, it fully recompenses your trouble, by opening a fine, clear, and extensive prospect.

The New Dictionary of Thoughts (p. 397)

The study of mathematics cultivates the reason; that of the languages, at the same time, the reason and the taste. The former gives the grasp and power to the mind; the latter both power and flexibility. The former by itself, would prepare us for a state of certainties, which nowhere exists; the latter, for a state of probabilities which is that of common life. Each, by itself, does but an imperfect work: in the union of both, is the best discipline for the mind, and the best mental training for the world as it is.

The New Dictionary of Thoughts (p. 397)

Einstein, Albert
But the creative principle resides in mathematics. In a certain sense, therefore, I hold true that pure thought can grasp reality, as the ancients dreamed.

Quoted in Heinz R. Pagels
The Cosmic Code (p. 40)

God does not care about our mathematical difficulties; He integrates empirically.

Quoted in Leopold Infeld
Quest—An Autobiography (p. 279)

But there is another reason for the high repute of mathematics: it is mathematics that offers the exact natural sciences a certain measure of security which, without mathematics, they could not attain.

Quoted in E.T. Bell
Men of Mathematics (p. xvi)

One reason why mathematics enjoys special esteem, above all other sciences, is that its laws are absolutely certain and indisputable, while those of all other sciences are to some extent debatable and in constant danger of being overthrown by newly discovered facts.

Sidelights on Relativity (p. 27)

As far as the laws of mathematics refer to reality, they are not certain; and as far as they are certain, they do not refer to reality.

Sidelights on Relativity (p. 28)

How can it be that mathematics, being after all a product of human thought which is independent of experience, is so admirably appropriate to the objects of reality? Is human reason, then, without experience, merely by taking thought, able to fathom the properties of real things?

Sidelights on Relativity (p. 28)

Mathematics are well and good but nature keeps dragging us around by the nose.

Quoted in A.P. French
Einstein: A Centenary Volume (p. 113)

Don't worry about your difficulties in mathematics; I can assure you that mine are still greater.

Source unknown

Ellis, Havelock
If mathematics were the only path to science . . . Nature would have been illegible for Goethe . . .

The Dance of Life
The Art of Thinking (pp. 129–30)

The mathematician's best work is art . . . a high and perfect art, as daring as the most secret dreams of imagination, clear and limpid. Mathematical genius and artistic genius touch each other.

Gustav Mittag-Lefler
Quoted in Havelock Ellis
The Dance of Life
The Art of Thinking (p. 139)

Here, we reach the sphere of mathematics, we are among processes which seem to some the most inhuman of all human activities and the most remote from poetry.

The Dance of Life
The Art of Thinking (pp. 138–9)

The mathematician has reached the highest rung on the ladder of human thought.

The Dance of Life
The Art of Thinking (p. 140)

Erath, V.
God is a child; and when he began to play, he cultivated mathematics. It is the most godly of man's games.

Das Blinde Spiel (p. 253)

Everett, Edward
In the pure mathematics we contemplate absolute truths which existed in the divine mind before the morning stars sang together, and which will continue to exist there when the last of their radiant host shall have fallen from heaven.

Quoted in E.T. Bell
The Queen of the Sciences (p. 20)

Eves, Howard W.
Mathematics may be likened to a large rock whose interior composition we wish to examine. The older mathematicians appear as persevering stone cutters slowly attempting to demolish the rock from the outside with hammers and chisel. The later mathematicians resemble expert miners who seek vulnerable veins, drill into these strategic places, and then blast the rock apart with well placed internal charges.

In Mathematical Circles
Volume 2
188 (p. 7)

Fabing, Harold
Mar, Ray
It is unsafe to talk mathematics. Folks don't understand.

Fischerisms (p. 3)

The pure mathematician starts with an unknown and ends with an unknown.

Fischerisms (p. 40)

Fairbairn, A.M.
The mathematics which have controlled and guided the Builder of the heavens, are identical with the mathematics which the astronomer in his study deduces from the idea of space given his own thoughts, and which he proves by the processes of his own reason.

The Philosophy of the Christian Religion (p. 37)

Feynman, Richard P.
Now you may ask, "What is mathematics doing in a physics lecture?" We have several possible excuses: first, of course, mathematics is an important tool, but that would only excuse us for giving the formula in two minutes. On the other hand, in theoretical physics we discover that all our laws can be written in mathematical form; and that this has a certain simplicity and beauty about it. So, ultimately, in order to understand nature it may be necessary to have a deeper understanding of mathematical relationships. But the real reason is that the subject is enjoyable, and although we humans cut nature up in different ways, and we have different courses in different departments, such compartmentalization is really artificial, and we should take our intellectual pleasures where we find them.

The Feynman Lectures on Physics
Volume I
Algebra

To those who do not know mathematics it is difficult to get across a real feeling as to the beauty, the deepest beauty, of nature . . . If you want to learn about nature, to appreciate nature, it is necessary to understand the language that she speaks in.

The Character of Physical Law (Chapter 2, p. 58)

Finn, Huckleberry
I had been to school . . . and could say the multiplication table up to six times seven is thirty-five, and I don't reckon I could ever get any further than that if I was to live forever. I don't take no stock in mathematics, anyway.

Quoted in Mark Twain
The Adventures of Huckleberry Finn
Chapter IV

Fitch, G.D.

Pure mathematics is a collection of hypothetical, deductive theories, each consisting of a definite system of primitive, *undefined* concepts or symbols and primitive, *unproved*, but self-consistent assumptions (commonly called axioms) together with their logically deducible consequences following by rigidly deductive processes without appeal to intuition.

In Henry P. Manning
The Fourth Dimension Simply Explained
Non-Euclidean Geometry of the Fourth Dimension (p. 58)

Forsyth, A.R.

Mathematics is one of the oldest of the sciences; it is also one of the most active, for its strength is the vigour of perpetual youth.

Nature
Presidential Address, British Association for the Advancement of Science
Section A, Volume 56, 1897 (p. 378)

Fourier, [Jean Baptiste] Joseph

. . . mathematical analysis is as extensive as nature itself.

Analytical Theory of Heat
Preliminary Discourse

The profound study of nature is the most fertile source of mathematical discoveries.

Quoted in Morris Kline
Mathematical Thought from Ancient to Modern Times (p. 671)

Franklin, W.S.

If a healthy minded person takes an interest in science, he gets busy with his mathematics and haunts the laboratory.

Quoted in Henry Crew
General Physics (p. 54)

Funk, Wilfred

The high wheel of heaven turns
With ten million shining urns,
A transcendent arc of light
In the interstellar night.
Flaming suns with fiery veils
Balance in their burning scales,
Spin their curves, and shining come
By a radiant rule of thumb.
Starry patterns, starry laws,
Buried in the primal cause,
Marshall to the cold decree
Of a bright geometry.

The Mathematics of the sky
Is frightening, when such as I
Cannot with impunity
Resolve the simple rule of three.

The Mathematics Teacher
Higher Mathematics
Volume XXIX, Number 1
January 1936 (p. 5)

Gassendi, P.

If we know anything we know it by mathematics; but those people have no concern for the true and legitimate science of things! they cling to trivialities!

Exercitationes paradoxicae adversus Aristotelicos
Problem I
Quoted in Pierre Duhem
The Aim and Structure of Physical Theory (p. 121)

Gauss, Carl Friedrich

Mathematics is the Queen of the Sciences, and Arithmetic the Queen of Mathematics.

Quoted in E.T. Bell
Men of Mathematics (p. xv)

Gibbon, Edward

The mathematics are distinguished by a peculiar privilege, that is, in the course of ages, they may always advance and can never recede.

The Decline and Fall of the Roman Empire
Chapter LII, section 59

Gibbs, J. Willard

Maxwell's Treatise on Electricity and Magnetism has done so much to familiarize students of physics with quaternion notations that it seems impossible that this subject should ever again be entirely divorced from the methods of multiple algebra.

I wish that I could say as much of astronomy. It is, I think, to be regretted that the oldest of the scientific applications of mathematics, the most dignified, the most conservative, should keep so far aloof from the youngest of mathematical methods.

Quoted in Florian Cajori
The Teaching and History of Mathematics in the United States (p. 157)

Gilbert, W.S.
Sullivan, Arthur

I'm very well acquainted too with matters mathematical.

The Mikado, and Other Plays
The Pirates of Penzance
Act I

Glanvill, Joseph
And for *Mathematical Sciences*, he that doubts their certainty, hath need of a dose of *Hellebore*.

The Vanity of Dogmatizing
Chapter XXI (p. 209)

. . . the knowledge we have of the *Mathematicks*, hath no reason to elate us; since by them we know but *numbers*, and *figures*, creatures of our own, and are yet ignorant of our *Maker's*.

The Vanity of Dogmatizing
Chapter XXI (pp. 209–10)

Gleason, Andrew
Like many great temples of some religions, mathematics may be viewed only from the outside by those uninitiated into its mysteries . . .

Science
Evolution of an Active Mathematical Theory
Volume 145, 31 July 1964 (p. 457)

If we do run into a paradox, we can probably save the structure of mathematics by patching it.

Quoted in Bryan H. Bunch
Mathematical Fallacies and Paradoxes (p. 110)

Goethe, Johann Wolfgang von
I have heard myself accused of being an opponent, an enemy of mathematics, which no one can value more highly than I, for it accomplishes the very thing whose achievement has been denied me.

Quoted in E.T. Bell
Men of Mathematics (p. xv)

Mathematics is like dialectics, an organ of the inner higher mind. Practising it is an art similar to eloquence. In both, *nothing is of value but the form*; towards the contents the users are indifferent. Whether mathematics calculates pennies or guineas, and whether rhetoric defends the true or the false, is quite outside the sphere of their interests.

Quoted in Walter R. Fuchs
Mathematics for the Modern Mind (p. 24)

The saying that no one who is unacquainted with, or a stranger to, geometry, should enter the school of the philosopher, does not mean that a man must be a mathematician in order to become a sage.

Criticisms, Reflections, and Maxims of Goethe (p. 207)

The mathematician is perfect only in so far as he is a perfect being, in so far as he perceives the beauty of truth; only then will his work be thorough, transparent, comprehensive, pure, clear, attractive, and even elegant.

> Quoted in JoAnne S. Growney
> *The American Mathematical Monthly*
> Are Mathematics and Poetry Fundamentally Similar?
> Volume 99, Number 2, February 1992 (p. 131)

A thorough advocate in a just cause, a penetrating mathematician facing the starry heavens, both alike bear the semblance of divinity.

> Quoted in JoAnne S. Growney
> *The American Mathematical Monthly*
> Are Mathematics and Poetry Fundamentally Similar?
> Volume 99, Number 2, February 1992 (p. 131)

The saying that no one who is unacquainted with, or a stranger to, geometry, should enter the school of the philosopher, does not mean that a man must be a mathematician in order to become a sage.

> *Criticism, Reflections, and Maxims of Goethe* (p. 207)

Graham, L.A.
Sing a song of sixpence—
  A mathman full of rye.
Four times twenty square feet
  Multiplied by π
Gives the total ground he covers
  While weaving an ellipse;
His path would have no area,
  If he had no nips.

> *Ingenious Mathematical Problems and Methods*
> Mathematical Nursery Rhyme No. 7

Graham, Ronald L.
Knuth, Donald E.
Patashnik, Oren
The ultimate goal of mathematics is to eliminate all need for intelligent thought.

> *Concrete Mathematics* (p. 56)

Guillen, Michael
. . . mathematics is not a science—it is not capable of proving or disproving the existence of real things. In fact, a mathematician's ultimate concern is that his or her inventions be logical, not realistic.

> *Bridges to Infinity* (p. 4)

Guruprasad, Venkata
Math is a perfect expression, like ballet or shaolin martial art.

> Quoted in Clifford A. Pickover
> *Keys to Infinity* (p. 147)

Guy, Richard K.
Mathematics often owes more to those who ask questions than to those who answer them. The solution of a problem may stifle interest in the area around it. But "Fermat's Last Theorem", because it is not yet a theorem, has generated a great deal of "good" mathematics, whether goodness is judged by beauty, by depth or by applicability.

> *Unsolved Problems in Number Theory* (p. v)

Halmos, Paul
Mathematics is abstract thought, mathematics is pure logic, mathematics is creative art. All these statements are wrong, but they are all a little right, and they are all nearer the mark than "mathematics is number" or "mathematics is geometric shapes." For the professional pure mathematician, mathematics is the logical dovetailing of a carefully selected sparse set of assumptions along with their surprising conclusions via a conceptually elegant proof. Simplicity, intricacy, and above all, logical analysis are the hallmark of mathematics.

> *American Scientist*
> Mathematics as a Creative Art
> Volume 56, Winter 1968 (p. 380)

Mathematics—this may surprise you or shock you some—is never deductive in its creation. The mathematician at work makes vague guesses, visualizes broad generalizations, and jumps to unwarranted conclusions. He arranges and rearranges his ideas, and he becomes convinced of their truth long before he can write down a logical proof. The conviction is not likely to come early—it usually comes after many attempts, many failures, many discouragements, many false starts. It often happens that months of work result in the proof that the method of attack they were based on cannot possibly work, and the process of guessing, visualizing and conclusion-jumping begins again.

> *American Scientist*
> Mathematics as a Creative Art
> Volume 56, Winter 1968 (p. 380)

The only way to learn mathematics is to do mathematics.

> *Hilbert Space Problem Book*
> Preface (p. vii)

Pure mathematics can be practically useful and applied mathematics can be artistically elegant.

> Quoted in Lynn Arthur Steen
> *Mathematics Tomorrow* (p. 12)

Applied mathematics cannot get along without pure, as an anteater cannot get along without ants, but not necessarily the reverse.

> Quoted in Lynn Arthur Steen
> *Mathematics Tomorrow* (p. 17)

Halsted, George Bruce
. . . mathematics, that giant pincers of scientific logic . . .

> *Science*
> Biology and Mathematics
> Volume XXII, Number 554, Friday, August 11, 1905 (p. 162)

Hankel, Hermann
In most sciences one generation tears down what another has built and what one has established another undoes. In mathematics alone each generation adds a new story to the old structure.

> Quoted in Morris Kline
> *Mathematical Thought from Ancient to Modern Times* (p. 200)

Hardy, Godfrey Harold
Beauty is the first test: there is no permanent place in the world for ugly mathematics.

> *A Mathematician's Apology*
> Chapter 10 (p. 85)

It is undeniable that a good deal of elementary mathematics—and I use the word 'elementary' in the sense in which professional mathematicians use it, in which it includes, for example, a fair working knowledge of the differential and integral calculus—has considerable practical utility. These parts of mathematics are, on the whole, rather dull; they are the parts which have the least aesthetic value. The 'real' mathematics of the 'real' mathematicians, the mathematics of Fermat and Euler and Gauss and Abel and Riemann, is almost wholly 'useless' (and this is as true of 'applied' as of 'pure' mathematics). It is not possible to justify the life of any genuine professional mathematician on the ground of the 'utility' of his work.

> *A Mathematician's Apology*
> Chapter 21 (pp. 119–20)

Harris, William Toney
Mathematics in its pure form, as arithmetic, algebra, geometry, and the applications of the analytic method, as well as mathematics applied to

matter and force, or statics and dynamics, furnishes us the peculiar study that gives to us, whether as children or as men, the command of Nature in this its quantitative aspect. Mathematics furnishes the instrument, the tool of thought which we wield in this realm.

Psychologic Foundations of Education
(pp. 325–6, fn (a))

Hauffman, Paul
That Mathematics could be a jewel may come as a surprise to those of us who struggled with multiplication tables as kids and now need help completing W-4 forms.

The Atlantic Magazine
The Man Who Loves Only Numbers
Volume 260, Number 5, November, 1987 (p. 66)

. . . mathematics is order and beauty at its purest, order that transcends the physical world.

The Atlantic Magazine
The Man Who Loves Only Numbers
Volume 260, Number 5, November, 1987 (p. 66)

Heinlein, Robert A.
Anyone who cannot cope with mathematics is not fully human. At best he is a tolerable subhuman who has learned to wear shoes, bathe, and not make messes in the house.

Time Enough for Love (p. 263)

. . . mathematics can never prove anything. No mathematics has any content. All any mathematics can do is—sometimes—turn out to be useful in describing some aspects of our so-called 'physical universe'. That is a bonus; most forms of mathematics are as meaning-free as chess.

The Number of the Beast
Chapter V (p. 39)

Heller, Joseph
She was a crazy mathematics major from the Wharton School of Business who could not count to twenty-eight each month without getting into trouble.

Catch-22
Chapter 8 (p. 72)

Helmholtz, Hermann von
Mathematics and music! the most glaring possible opposites of human thought! and yet connected, mutually sustained!

Popular Lectures on Scientific Subjects
On the Physiological Causes of Harmony in Music (p. 62)

Henry, O.
His mathematics carried with it a momentary qualm and a lesson. The thought had not occurred to him that the thought could possibly occur to me not to ride at his side on that red road to revenge and justice. It was the higher calculus. I was booked for the trail. I began to eat more beans.

Tales of O. Henry
A Technical Error

Herbart, Johann Friedrich
The idea that aptitude for mathematics is rarer than aptitude for other subjects is merely an illusion which is caused by belated or neglected beginners.

Quoted in Stanley Gudder
A Mathematical Journey (p. ix)

Herschel, Sir John
Admission to its sanctuary, and to the privileges and feelings of a votary, is only to be gained by one means—*sound and sufficient knowledge of mathematics, the greatest instrument of all exact inquiry, without which no man can ever make such advances in this or any other of the higher departments of science as can entitle him to form an independent opinion on any subject of discussion within their range.*

Outlines of Astronomy
Part I
Introduction (p. 26)

Hersh, Reuben
Formalized mathematics, to which most philosophizing has been devoted in recent years, is in fact hardly to be found anywhere on earth or in heaven outside the texts and journals of symbolic logic.

Quoted in John D. Barrow
Pi in the Sky (p. 140)

Hertz, Heinrich
One cannot escape the feeling that these mathematical formulas have an independent existence and an intelligence of their own, that they are wiser than we are, wiser even than their discoverers, that we get more out of them than was originally put into them.

Quoted in Morris Kline
Mathematics and the Search for Knowledge (p. 144)

Hilbert, David
For it is true . . . that mathematics is not, generally speaking, a popular subject.

Quoted in Constance Reid
Hilbert (p. 154)

We hear within us the perpetual call: There is the problem. Seek its solution. You can find it by pure reason, for in mathematics there is no *ignorabimus*.

Bulletin of the American Mathematical Society
Mathematical Problems
Volume 8, 2nd series, October 1901–July 1902 (p. 445)

A mathematical problem should be difficult in order to entice us, yet not completely inaccessible, lest it mock at our efforts.

Quoted in David M. Burton
The History of Mathematics (p. 289)

Mathematics . . . is nothing more than a game played according to certain simple rules with meaningless marks on a paper.

Quoted in E.T. Bell
Mathematics: Queen and Servant of Science
Mathematical Truth (p. 21)

Hill, Thomas
The mathematics are usually considered as being the very antipodes of Poesy. Yet Mathesis and Poesy are of the closest kindred, for they are both works of the imagination.

North American Review
The Imagination in Mathematics
Volume 85, July 1857 (p. 229)

Mathematics and Poetry are . . . the utterance of the same power of imagination, only that in the one case it is addressed to the head, in the other, to the heart.

North American Review
The Imagination in Mathematics
Volume 85, July 1857 (p. 230)

Hodges, Wilfrid
Mathematics is not a topic that one can easily approach with a virgin mind.

Building Models by Games (p. 1)

Hogben, Lancelot
If mathematics is a game, there is no reason why people should play it if they do not want to. With football, it belongs to those amusements without which life would be endurable.

Mathematics for the Million (p. 22)

If the rules of mathematics are rules of grammar, there is no stupidity involved when we fail to see that a mathematical truth is obvious.

Mathematics for the Million (p. 22)

Holmes, Oliver Wendell

There is no elasticity in a mathematical fact; if you bring up against it, it never yields a hair's breadth; everything must go to pieces that comes in collision with it. What the mathematician knows being absolute, unconditional, incapable of suffering question, it should tend, in the nature of things, to breed a despotic way of thinking. So of those who deal with the palpable and often unmistaken facts of external nature; only to a less degree. Every probability—and most of our common, working beliefs are probabilities—is provided with *buffers* at both ends, which break the force of opposite opinions clashing against it . . .

The Autocrat of the Breakfast-Table
Chapter 3

Holmes, Oliver Wendell, Jr.

The law embodies the story of a nation's development through many centuries, and it cannot be dealt with as if it contained only the axioms and corollaries of a book of mathematics.

The Common Law
Lecture I (p. 1)

Holmes, Sherlock

While the individual man is an insoluble puzzle, in the aggregate he becomes a mathematical certainty. You can, for example, never foretell what any one man will be up to, but you can say with precision what an average number will be up to. Individuals vary, but percentages remain constant. So says the statistician.

In Arthur Conan Doyle's
The Complete Sherlock Holmes
The Sign of the Four

Honsberger, Ross

Mathematics abounds in bright ideas. No matter how long and hard one pursues her, mathematics never seems to run out of exciting surprises. And by no means are these gems to be found only in difficult work at an advanced level. All kinds of simple notions are full of ingenuity.

Mathematical Morsels (p. vii)

Hubbard, John

The Mathematicks too our tho'ts employ,
Which nobly elevate the Student's joy:
The little Euclids round the table set
And at their rigid demonstrations sweat.

Quoted in Florian Cajori
The Teaching and History of Mathematics in the United States (p. 30)

Hudson, Hilda Phoebe

To all of us who hold the Christian belief that God is truth, anything that is true is a fact about God, and mathematics is a branch of theology.

Quoted in Eric T. Bell
The Magic of Numbers (p. 385)

But however we think of heaven, it is hard to imagine astronomy and botany surviving as they are, and having much interest or importance there . . . On the other hand it is just as hard to imagine pure mathematics not surviving. The laws of thought, and especially of number, must hold good in heaven, whether it is a place or a state of mind; for they are independent of any particular sphere of existence, essential to Being itself, to God's being as well as ours, laws of His mind before we learned them. The multiplication table will hold good in heaven . . .

Quoted in Eric T. Bell
The Magic of Numbers (pp. 385–6)

The two main divisions of mathematics, analysis and geometry, correspond with some exactness to the two great mysteries of the Christian faith, the Trinity and the Incarnation.

Quoted in Eric T. Bell
The Magic of Numbers (p. 386)

Huntley, Henry Edwards

To the aesthetically minded mathematician much mathematics reads like poetry.

The Divine Proportion: a Study in Mathematical Beauty (p. vii)

If you . . . contemplate a career in pure or applied mathematics, whether in industry or research, or in the teaching profession, you should be warned that although there can be one infallible, enduring reward for you in this pursuit—joy in creative activity—there stand certain discouraging hazards, of which four may be noted briefly:

1. The burden of hard mental concentration is a *sine qua non*. You may find that you have to live with a problem day and night for weeks, giving all you have of mental resources in order to solve it: no inspiration without perspiration.

2. Your best efforts may be fruitless. Despite extravagant expenditure of time and skill, the result is nil. Disappointment, frustration and near-despair are common experiences of serious mathematicians.

3. You may be lonely. Scarcely anyone will appreciate your work because few will be capable of understanding it.

4. The results you do obtain will always appear to be disproportionately meager in comparison with the effort you expended to produce them: "The mountain laboured and brought forth a mouse."

The one sure path to satisfaction in a mathematics career is to cultivate assiduously the aesthetic appreciation of the discipline. That pleasure will not fade, it will grow with exercise.

The Divine Proportion: a Study in Mathematical Beauty (p. 3)

Huxley, Aldous
I admit that mathematical science is a good thing. But excessive devotion to it is a bad thing.

Quoted in James R. Newman
The World of Mathematics
Volume IV (p. 2027)

Huxley, Thomas H.
[Mathematics] is that [subject] which knows nothing of observation, nothing of experiment, nothing of induction, nothing of causation!

Lay Sermons, Addresses and Reviews (p. 168)

Mathematics may be compared to a mill of exquisite workmanship which grinds you stuff of any degree of fineness; but, nevertheless, what you get out depends on what you put in; and as the grandest mill in the world will not extract wheat-flour from peas-cods, so pages of formulae will not get a definite result out of loose data.

Collected Essays
Geological Reform

Issigoinis, Sir Alec
All creative people hate mathematics. It's the most uncreative subject you can study.

Quoted in
The Australian
5 October 1988

Jacobi, Karl G.J.
Mathematics exists solely for the honour of the human mind.

Quoted in Walter R. Fuchs
Mathematics for the Modern Mind (p. 24)

Jeans, Sir James Hopwood
. . . nature seems very conversant with the rules of pure mathematics, as our mathematicians have formulated them in their studies, out of their own inner consciousness and without drawing to any appreciable extent on their experience of the outer world.

The Mysterious Universe
Into Deep Waters (p. 154)

Jeffers, Robinson
Science and mathematics
Run parallel to reality, they symbolize it, they squint at it,
They never touch it . . .

The Beginning & the End
The Silent Shepherds

Kac, Mark
. . . there are those who believe that mathematics can sustain itself and
grow without any further contact with anything outside itself, and those
who believe that nature is still and always will be one of the main (if
not the main) sources of mathematical inspiration. The first group is
identified as "pure mathematicians" (though "purist" would be more
adequate) while the second is, with equal inadequacy, referred to as
"applied".

Quoted in Robert W. Ritchie (Editor)
New Directions in Mathematics (p. 60)

Kant, Immanuel
Mathematical judgments are always synthetical.

The Critique of Pure Reason
Introduction, Section V

But mathematics, certainly, does not play the smallest part in the charm
and movement of the mind produced by music. Rather is it only the
indispensable condition (*conditio sine qua non*) of that proportion of the
combining as well as changing impressions which makes it possible to
grasp them all in one and prevent them from destroying one another,
and to let them, rather, conspire towards the production of a continuous
movement and quickening of the mind by affections that are in unison
with it, and thus towards a serene self-enjoyment.

The Critique of Judgment
Critique of Aesthetic Judgment
Section 53

Pure mathematics can never deal with the real existence of things, but
only with their possibility, that is to say, with the possibility of an
intuition answering to the conceptions of the things. Hence it cannot
touch the question of cause and effect, and consequently, all the finality
there observed must always be regarded simply as formal, and never as
a physical end.

The Critique of Judgment
Critique of Teleological Judgment
Section 63, fn 1

In every department of physical science there is only so much science, properly so-called, as there is mathematics.

Quoted in Morris Kline
Mathematics and the Physical World (p. vii)

Karpinski, L.C.

As a professor of mathematics I am practically required by the ethics of the profession to be absent-minded, unmethodical, and inconsistent in many ways fatal to bibliographical excellence.

Bibliography of Mathematical Works Printed in America through 1850
Preface (p. viii)

Kasner, Edward
Newman, James R.

Here, then, in mathematics we have a universal language, valid, useful, intelligible everywhere in place and in time—in banks and insurance companies, on the parchments of the architects who raised the Temple of Solomon, and on the blueprints of the engineers who, with their calculus of chaos, master the winds. Here is a discipline of a hundred branches, fabulously rich, literally without limit in its sphere of application, laden with honors for an unbroken record of magnificent accomplishment. Here is a creation of the mind, both mystic and pragmatic in appeal. Austere and imperious as logic, it is still sufficiently sensitive and flexible to meet each new need.

Mathematics and the Imagination (p. 358)

Mathematics is an activity governed by the same rules imposed upon the symphonies of Beethoven, the paintings of DaVinci, and the poetry of Homer. Just as scales, as the laws of perspective, as the rules of metre seem to lack fire, the formal rules of mathematics may appear to be without lustre. Yet ultimately, mathematics reaches pinnacles as high as those attained by the imagination in its most daring reconnoiters. And this conceals, perhaps, the ultimate paradox of science. For in their prosaic plodding both logic and mathematics often outstrip their advance guard and show that the world of pure reason is stranger than the world of pure fancy.

Mathematics and the Imagination (p. 362)

Keller, Helen

Now I feel as if I should succeed in doing something in mathematics, although I cannot see why it is so very important . . . The knowledge doesn't make life any sweeter or happier, does it?

The Story of My Life
Letter to Mrs. Laurence Hutton
May 29, 1898 (p. 242)

. . . I've said goodbye to Mathematics forever, and I assure you, I was delighted to see the last of those horrid goblins!

The Story of My Life
Letter to Mr. John D. Wright
December 9, 1900 (p. 270)

Kepler, Johannes
The Laws of Nature are but the mathematical thoughts of God.

The Mathematical Gazette
Mathematical Clubs in School
Volume 12, Number 176, May 1925 (p. 389)

Keyser, Cassius J.
But for the unattainable ideal of logical perfection, we should be without the miracles of modern Mathematics.

Mole Philosophy and Other Essays (p. 3)

Every major concern among the intellectual concerns of man is a concern of mathematics.

Mole Philosophy and Other Essays (p. 93)

It is customary to speak of mathematics, of pure mathematics, and of applied mathematics, as if the first were a *genus* owning the other two as *species*. The custom is unfortunate because it is misleading.

Mole Philosophy and Other Essays (pp. 109–10)

It can, you see, be said, with the same approximation to truth, that the whole of science, including mathematics, consists in the study of transformations or in the study of relations.

Mathematical Philosophy
Chapter X (p. 168)

Mathematics thus belongs to the great family of spiritual enterprises of man. These enterprises, all the members of the great family, however diverse in form in modes of life, in methods of toil, in their progress along the way that leads toward logical rectitude, are alike children of one great passion. In genesis, in spirit and aspiration, in motive and aim, mutual science, theology, philosophy, jurisprudence, religion and art are one with mathematics; they are all of them sprung from the human spirit's craving for invariant reality in a world of tragic change; they all of them aim at rescuing man from "the blind hurry of the universe from vanity to vanity" [Bertrand Russell]: they seek cosmic stability—a world of abiding worth, where the broken promises of hope shall be healed and the infinite aspiration shall cease to be mocked.

The Human Worth of Rigorous Thinking (p. 48

Mathematics, like any other cardinal activity of the human spirit, has an individuality of its own.

Mathematics and the Question of Cosmic Mind with Other Essays
The Meaning of Mathematics (p. 3)

King, Jerry P.

. . . pure mathematics is mathematics for mathematics' sake and applied mathematics is mathematics for something else.

The Art of Mathematics
Chapter 2 (p. 26)

Klein, Felix

But it should always be required that a mathematical subject not be considered exhausted until it has become intuitively evident . . .

Quoted in Morris Kline
Mathematical Thought from Ancient to Modern Times (p. 904)

Mathematics in general is fundamentally the science of self evident things.

Mathematics: Queen and Servant of Science
Mathematical Truth (pp. 19–20)

The mathematics of our day appears to me like a large weapons shop in peace time. The store window is filled with show pieces whose ingenious, artful and pleasing design enchants the connoisseur. The real origin and purpose of these things, to attack and defeat the enemy, has retreated so far into the background of consciousness as to be forgotten.

Development of Mathematics in the 19th Century (p. 65)

In fact, mathematics has grown like a tree, which does not start at its tiniest rootlets and grow merely upward, but rather sends its roots deeper and deeper at the same time and rate that its branches and leaves are spreading upward . . . *We see, then, that as regards the fundamental investigation in mathematics, there is no final ending, and therefore on the other hand, no first beginning* . . .

Elementary Mathematics from an Advanced Standpoint (p. 15)

The science of mathematics may be compared to a tree thrusting its roots deeper and deeper into the earth and freely spreading out its shady branches to air.

Bulletin of the American Mathematical Society
Arithmetizing of Mathematics
Volume 11, Number 8, 2nd series, May 1896 (p. 248)

Kline, Morris

Perhaps the best reason for regarding mathematics as an art is not so much that it affords an outlet for creative activity as that it provides spiritual values. It puts man in touch with the highest aspirations and loftiest goals. It offers intellectual delight and the exaltation of resolving the mysteries of the universe.

Mathematics: a Cultural Approach (p. 671)

Mathematics may be the queen of the sciences and therefore entitled to royal prerogatives, but the queen who loses touch with her subjects may lose support and even be deprived of her realm. Mathematicians may like to rise into the clouds of abstract thought, but they should, and indeed they must, return to earth for nourishing food or else die of mental starvation. They are on safer and saner ground when they stay close to nature. As Wordsworth put it, "Wisdom oft is nearer when we stoop than when we soar."

Mathematics and the Physical World (p. 537)

. . . mathematics has determined the direction and content of much philosophic thought, has destroyed and rebuilt religious doctrines, has supplied substance to economic and political theories, has fashioned major painting, musical, architectural, and literary styles, has fathered our logic, and has furnished the best answers we have to fundamental questions about the nature of man and his universe. As the embodiment and most powerful advocate of the rational spirit, mathematics has invaded domains ruled by authority, custom, and habit, and supplanted them as the arbiter of thought and action. Finally, as an incomparably fine human achievement mathematics offers satisfactions and aesthetic values at least equal to those offered by any other branch of our culture.

Mathematics in Western Culture (p. ix)

When we consider the number of fields on which mathematics impinges and the number of these over which it already gives us mastery or partial mastery, we are tempted to call it a method of approach to the universe of physical, mental, and emotional experiences. It is distillation of highest purity that exact thought has extracted from man's efforts to understand nature, to impart order to the confusion of events occurring in the physical world, to create beauty, and to satisfy the natural proclivity of the healthy brain to exercise itself.

Mathematics in Western Culture (pp. 471–2)

The conquest of new domains of mathematics proceeds somewhat as do military conquests. Bold dashes into enemy territory capture strongholds. These incursions must then be followed up and supported by broader, more thorough and more cautious operations to secure what has been only tentatively and insecurely grasped.

Mathematical Thought from Ancient to Modern Times
Chapter 19 (p. 400)

We should drop the ideas that mathematics and what mathematics says about the world are indubitable truths. Today there is no agreement among mathematicians on fundamental principles . . . Mathematics is not the universally accepted, precise body of knowledge that it was thought to be 100 years ago when scholars believed that it revealed the design of the universe.

U.S. News and World Report
Mathematics: From Precision to Doubt in 100 years
January 26, 1981

Lakatos, Imre
Mathematics has been trivialized, derived from indubitable, trivial axioms in which only absolutely clear trivial terms figure, and from which truth pours down in clear channels.

Mathematics, Science and Epistemology
Volume 2 (p. 10)

Mathematics does not grow through a monotonous increase of the number of indubitably established theorems, but through the incessant improvement of guesses by speculation and criticism.

Quoted in Michael Guillen
Bridges to Infinity (p. 19)

Lamb, Sir Horace
A traveler who refuses to pass over a bridge until he has personally tested the soundness of every part of it is not likely to go far; something must be risked, even in mathematics.

Quoted in Morris Kline
Mathematical Thought from Ancient to Modern Times (p. 468)

Langer, R.E.
Rich in its past, dynamic in the present, prodigious for the future, replete with simple and yet profound ideas and methods, surely mathematics can give something to anyone's culture.

American Mathematical Monthly
The Things I Should Have Done, I Did Not Do
Volume 59, September 1952 (p. 445)

Lasserre, François
Ask a philosopher 'What is philosophy?' or a historian 'What is history?' and they will have no difficulty in giving an answer. Neither of them,

in fact, can pursue his own discipline without knowing what he is searching for. But ask a mathematician 'What is mathematics?' and he may justifiably reply that he does not know the answer but that does not stop him from doing mathematics.

Quoted in John D. Barrow
Pi in the Sky (p. 1)

Lec, Stanislaw J.

I don't agree with mathematics; the sum total of zeros is a frightening figure.

More Unkempt Thoughts (p. 26)

Lehrer, Tom

In one word he told me the secret of success in mathematics: Plagiarize . . . only be sure always to call it please research.

Lobachevsky
(Satirical song)

Counting sheep
When you're trying to sleep,
Being fair
When there's something to share,
Being neat
When you're folding a sheet,
That's mathematics!
When a ball
Bounces off of a wall,
When you cook
From a recipe book,
When you know
How much money you owe,
That's mathematics!

Source unknown
(To the tune of *"That's Entertainment"*)

Lem, Stanislaw

". . . Let's have a love poem, lyrical, pastoral, and expressed in the language of pure mathematics. Tensor algebra mainly, with a little topology and higher calculus if need be. But with the feeling, you understand, and in the cybernetic spirit."

"Love and tensor algebra? Have you taken leave of your sense!" Trurl began, but stopped, for his electronic bard was already declaiming:

Come, let us hasten to a higher plane,
Where dyads tread the fairy fields of Venn,

Their indices bedecked from one to n,
Commingled in an endless Markov chain!

Come, every frustum longs to be a cone,
And every vector dreams of matrices.
Hark to the gentle gradient of the breeze:
It whispers of a more ergodic zone.

In Riemann, Hilbert or in Banach space
Let superscripts and subscripts go their ways.
Our asymptotes no longer out of phase,
We shall encounter, counting, face to face.

I'll grant thee random access to my heart,
Thou'lt tell me all the constants of thy love;
And so we two shall all love's lemmas prove,
And in our bound partition never part.

For what did Cauchy know, or Christoffel,
Or Fourier, or any Boole or Euler,
Wielding their compasses, their pens and rulers,
Of thy supernal sinusoidal spell?

Cancel me not—for what then shall remain?
Abscissas, some mantissas, modules, modes,
A root or two, a torus and a node:
The inverse of my verse, a null domain.

Ellipse of bliss, converge, O lips divine!
The product of our scalars is defined!
Cyberiad draws nigh, and the skew mind
Cuts capers like a happy haversine.

I see the eigenvalue in thine eye,
I hear the tender tensor in thy sigh.
Bernoulli would have been content to die,
Had he but known such $a^2 \cos 2\phi$!

The Cyberiad
The First Sally (p. 53)

Leslie, John
The study of mathematics holds forth two capital objectives; while it traces the beautiful relations of figure and quantity, it likewise accustoms the mind to the invaluable exercise of patient attention and accurate reasoning. Of these distinct objects the last is perhaps the most important in a course of liberal education. For this purpose, the geometry of the ancients is the most powerfully recommended, as bearing the stamp of that acute people, and displaying the finest specimens of logical deduction. Some of the propositions, indeed, might be reached by a sort

of algebraic calculation; but such an artificial mode of procedure gives only an apparent facility, and leaves no clear or permanent impression on the mind.

Elements of Geometry (pp. v–vi)

Libchaber, Albert

A physicist would ask me, How does this atom come here and stick there? And what is the sensitivity to the surface? And can you write the Hamiltonian of the system? And if I tell him, I don't care, what interests me is this shape, the mathematics of the shape and the evolution, the bifurcation from this shape to that shape to this shape, he will tell me, that's not physics, you are doing mathematics.

Quoted in J. Gleick
Chaos: Making a New Science (pp. 210–11)

Lieber, Lillian R.

When we learn to drive a car we are able to "go places" easily and pleasantly instead of walking to them with a great deal of effort. And so you will see that the more Mathematics we know the EASIER life becomes, for it is a TOOL with which we can accomplish things that we could not do at all with our bare hands. Thus Mathematics helps our brains and hands and feet, and can make a race of supermen out of us.

The Education of T.C. MITS (p. 45)

. . . the knowledge of Mathematics of the average college graduate stops with what was known about 300 years ago!

The Education of T.C. MITS (p. 125)

Lindsay, R.B.

Of one thing we may be sure: physics without mathematics will forever be incomprehensible.

Scientific Monthly
On the Relation of Mathematics and Physics
Volume 59, December 1944 (p. 460)

Lobachevskii, Nikolai Ivanovich

There is no branch of mathematics however abstract which may not some day be applied to phenomena of the real world.

Quoted in Stanley Gudder
A Mathematical Journey (p. 36)

Locke, John
Would you have a man reason well, you must use him to it betimes; exercise his mind observing the connection between ideas, and following them in train. Nothing does this better than mathematics, which therefore, I think should be taught to all who have the time and opportunity, not so much to make them mathematicians, as to make them reasonable creatures; for though we all call ourselves so, because we are born to it if we please, yet we may truly say that nature gives to us but the seeds of it, and we are carried no farther than industry and application have carried us.

Of the Conduct of the Understanding
Section 6

I have before mentioned mathematics, wherein algebra gives new helps and views to the understanding. If I propose these it is not to make every man a thorough mathematician or deep algebraist; but I think the study of them is of infinite use even to grown men; first by experimentally convincing them, that to make anyone reason well, it is not enough to have parts wherewith he is satisfied, and that serve him well enough in his ordinary course.

Of the Conduct of the Understanding
Section 7

Secondly, the study of mathematics would show them the necessity there is in reasoning, to separate all the distinct ideas, and to see the habitudes that all those concerned in the present inquiry have to one another, and to lay by those which relate not to the proposition in hand, and wholly to leave them out of the reckoning. This is that which, in other respects besides quantity is absolutely requisite to just reasoning, though in them it is not so easily observed and so carefully practiced.

Of the Conduct of the Understanding
Section 7

Lodge, Sir Oliver
An equation is the most serious and important thing in mathematics.

Easy Mathematics

Lucas, William F.
Although some older art, music or wines may be better than the newer, it is rather unlikely that this would often apply to science or mathematics.

Quoted in Lynn Arthur Steen
Mathematics Tomorrow (p. 65)

Mac Lane, Saunders
Mathematics, springing from the soil of basic human experience with numbers and data and space and motion, builds up a far-flung

architectural structure composed of theorems which reveal insights into the reasons behind appearances and of concepts which relate totally disparate concrete ideas.

American Mathematical Monthly
Of Course and Courses
Volume 61, March 1954 (p. 152)

Mach, Ernst

Strange as it may sound, the power of mathematics rests on its evasion of all unnecessary thought and on its wonderful saving of mental operations.

Quoted in E.T. Bell
Men of Mathematics (p. xvi)

Mathematics may be defined as the economy of counting. There is no problem in the whole of mathematics which cannot be solved by direct counting. But with the present implements of mathematics many operations of counting can be performed in a few minutes, which, without mathematics, would take a lifetime.

Quoted in J.W. Mellor
Higher Mathematics for Students of Chemistry and Physics (p. 184)

Mann, Lee

With *ordinal* intelligence any *irrational* person can teach the new modern math to the *least common multiple*. (Anyone with a forked tongue, two left thumbs, and a touch of insanity in the family will find it a natural.)

Arithmetic Teacher
The Digit it is!
Volume 13, Number 8, December 1966 (p. 66)

Mann, Thomas

I tell them that if they will occupy themselves with the study of mathematics they will find in it the best remedy against the lusts of the flesh.

The Magic Mountain
Choler and Worse (p. 417)

Mathematical Sciences Education Board

For most students, school mathematics is a habit of problem-solving without sense-making: one learns to read the problem, to extract the relevant numbers and the operation to be used, to perform the operation, and to write down the result—without ever thinking about what it all means.

Reshaping School Mathematics (p. 32)

Mencke, J.B.
[Mathematics] guides our minds in an orderly way, and furnishes us simple and rational principles by means of which ambiguities are clarified, disorder is converted into order, and complexities are analyzed into their component parts.

The Charlatanry of the Learned
Lecture II (p. 152)

[Mathematics] includes much that will neither hurt one who does not know it nor help one who does.

The Charlatanry of the Learned
Lecture II (p. 152)

Mendès, Michel
A talk in mathematics should be one of four things: beautiful, deep, surprising . . . or short.

Remark, c. 1986
Source unknown

Meyer, Walter
In a time when much of the world's geography has been explored, and space exploration is restricted to astronauts, mathematics offers fertile ground for exploring the unknown.

Humanistic Mathematics Network Journal
Missing Dimensions of Mathematics
No. 11, February 1995

Mill, John Stuart
The character of necessity ascribed to the truths of mathematics and even the peculiar certainty attributed to them is an illusion.

Quoted in Morris Kline
Mathematical Thought from Ancient to Modern Times (p. 861)

The peculiarity of the evidence of mathematical truths is that all the argument is on one side. There are no objections, and no answers to objections.

On Liberty
Chapter II (p. 44)

Milligan, Spike
MORIARTY: How are you at Mathematics?

HARRY SECOMBE: I speak it like a native.

The Goon Show, BBC Radio

Monboddo, Lord James Burnett
Those who have studied mathematics much, and no other science, are apt to grow so fond of them, as to believe that there is no certainty in any other science, nor any other axioms than those of Euclid.

Quoted in
Edinburgh Review
Volume 52
January 1836 (p. 248)

More, L.T.
The supreme value of mathematics to science is due to the fact that scientific laws and theories have their best, if not their only complete, expression in mathematical formulae: and the degree of accuracy with which we can express scientific theory in mathematical terms is a measure of the state of a science.

The Limitations of Science
Classical & New Mechanics (p. 150)

The goal of science is mathematics, and while mathematics may be said to be the only true science since it has the only true scientific method, mathematics is not a science because it deals with abstractions and ignores concrete phenomena.

The Limitations of Science
Classical & New Mechanics (p. 151)

Morse, Harold Marston
But mathematics is the sister, as well as the servant, of the arts and is touched with the same madness and genius.

Quoted in Stanley Gudder
A Mathematical Journey (p. 81)

Morton, Henry Vellam
In the dusk of a lane I met a shepherd with his sheep. A small dog with the expression of a professor of mathematics does all the work.

In Search of Scotland
Chapter 1, section 3 (p. 8)

Moultrie, John
There's nothing in the world (that is in Trinity)
To make us poets happy;—I detest
Your Hebrew, Greek and heathenish Latinity,
And Mathematics are a bore at best.

Poems
Sir Launfal
xii

Newman, James R.
There are two ways to teach mathematics. One is to take real pains toward creating understanding visual aids, that sort of thing. The other is the old British system of teaching until you're blue in the face.

New York Times
September 39, 1956

Nietzsche, Friedrich
... *mathematics*, which would certainly have not originated if it had been known from the beginning that there is no exactly straight line in nature, no real circle, no absolute measure.

Human, All-Too-Human
Section One, Number 11 (p. 19)

Oman, John
Beauty is a conspicuous element in the abstract completeness aimed at in the higher mathematics . . .

The Natural and the Supernatural
Value and Validity (p. 211)

Oppenheimer, Julius Robert
Today, it is not only that our kings do not know mathematics, but our philosophers do not know mathematics and—to go a step further—our mathematicians do not know mathematics.

Harper's Magazine
The Tree of Knowledge
Volume 217, Number 1301, October 1958 (p. 55)

Page, Ray
. . . Today our world of automation revolves around science and science in turn rests on mathematics.

Quote
Volume 53, Number 20, May 14, 1967 (p. 380)

Pascal, Blaise
But dull minds are never either intuitive or mathematical.

Pensées
Section I, 1

For it is to judgment that perception belongs, as science belongs to intellect. Intuition is the part of judgment, mathematics of intellect.

Pensées
Section I, 4

There is a great difference between the *spirit of Mathematics* and the *spirit of Observation*. In the former, the principles are palpable, but remote from common use; so that from want of custom it is not easy to turn our head in that direction; but if it be thus turned ever so little, the principles are seen fully confessed, and it would argue a mind incorrigibly false, to reason inconsequentially on principles so obtrusive, that it is hardly possible to overlook them.

Quoted in
Edinburgh Review
Volume 52
January 1836 (p. 241)

Peirce, Benjamin
Mathematics is the science which draws necessary conclusions.

American Journal of Mathematics
Linear Associative Algebra
Volume 4, 1881 (p. 97)

Mathematics is not the discoverer of laws, for it is not induction; neither is it the framer of theories, for it is not hypothesis; but it is the judge over both, and it is the arbiter to which each must refer its claims; and neither law can rule nor theory explain without the sanction of mathematics.

American Journal of Mathematics
Linear Associative Algebra
Volume 4, 1881 (p. 97)

Peirce, Charles Sanders
. . . metaphysics has always been the ape of mathematics.

The Monist
The Architecture of Theories
Volume 1, Number 2, January 1891 (p. 174)

Mathematics is the most abstract of all the sciences. For it makes no external observations, nor asserts anything as a real fact. When the mathematician deals with facts, they become for him mere "hypotheses"; for with their truth he refuses to concern himself. The whole science of mathematics is a science of hypotheses; so that nothing could be more completely abstracted from concrete reality.

The Monist
The Regenerated Logic
Volume 7, Number 1, October 1896 (p. 23)

Peterson, Ivars
To most outsiders, modern mathematics is unknown territory. Its borders are protected by dense thickets of technical terms; its landscapes are a mass of indecipherable equations and incomprehensible concepts. Few realize that the world of modern mathematics is rich with vivid images and provocative ideas.

The Mathematical Tourist
Preface (p. xiii)

. . . mystery is an inescapable ingredient of mathematics. Mathematics is full of unanswered questions, which far outnumber known theorems and results. It's the nature of mathematics to pose more problems than it can solve. Indeed, mathematics itself may be built on small islands of truth comprising the pieces of mathematics that can be validated by relatively short proofs. All else is speculation.

Islands of Truth: A Mathematical Mystery Cruise
Preface (p. xvi)

Pieri, Mario
Mathematics is the hypothetico-deductive science.

Quoted in Cassius J. Keyser
The Pastures of Wonder
The Realm of Mathematics (p. 24)

Pierpont, James
We who stand on the threshold of a new century can look back on an era of unparalleled progress. Looking into the future an equally bright prospect greets our eyes; on all sides fruitful fields of research invite our labor and promise easy and rich returns. Surely this is the golden age of mathematics!

Bulletin of the American Mathematical Society
2nd Series
The History of Mathematics in the Nineteenth Century
Volume 11, 1904–1905 (p. 159)

Poe, Edgar Allan
As poet *and* mathematician, he would reason well; as mere mathematician, he could not have reasoned at all, and thus would have been at the mercy of the Prefect.

Seven Tales
The Purloined Letter (p. 231)

Poiani, Eileen L.
Like it or not, mathematics opens career doors, so it's downright practical to be prepared.

Quoted in Lynn Arthur Steen
Mathematics Tomorrow (p. 158)

Mathematics plays the critical filter role not only at the college level, but also in the work force.

Quoted in Lynn Arthur Steen
Mathematics Tomorrow (p. 160)

Poincaré, Henri
How does it happen that there are people who do not understand mathematics? If the science invokes only the rules of logic, those accepted by all well-formed minds, if its evidence is founded on principles that are common to all men, how does it happen that there are so many people who are entirely impervious to it?

Science and Method
Mathematical Discovery (pp. 46–7)

It is necessary to add that mathematicians themselves are not infallible.

Science and Method
Mathematical Discovery (p. 47)

To the superficial observer scientific truth is unassailable, the logic of science is infallible; and if scientific men sometimes make mistakes, it is because they have not understood the rules of the game. Mathematical truths are derived from a few self-evident propositions, by a chain of flawless reasonings; they are imposed not only on us, but on Nature itself. By them the Creator is fettered, as it were, and His choice is limited to a relatively small number of solutions. A few experiments, therefore, will be sufficient to enable us to determine what choice He has made. From each experiment a number of consequences will follow by a series of mathematical deductions, and in this way each of them will reveal to us a corner of the universe. This, to the minds of most people, and to students who are getting their first ideas of physics, is the origin of certainty in science.

Science and Hypothesis
Author's Preface (p. xxi)

Mathematics has a triple end. It is to furnish an instrument for the study of nature. But that is not all. It has a philosophic end, and I dare say it, an esthetic end . . . Those skilled in mathematics find in it pleasure akin to those which painting and music give. They admire the delicate harmony of numbers and of forms; they marvel when a new discovery opens an unexpected perspective; and is this pleasure not esthetic, even though the senses have no part in it?

Sur les rapports de l'analyse pure et de la physique mathématique
Report Internat. Cong. Math.
Zürich 1897 (p. 82)

Poisson, Simeon
Life is good for only two things, discovering mathematics and teaching mathematics.

Mathematical Magazine
Filler
Volume 64, Number 1, February 1991 (p. 44)

Polanyi, Michael

While applied mathematics is object-directed, pure mathematics has no outside object; being concerned with objects of its own creation, it may be described as 'object creating'.

Personal Knowledge
Chapter 5 (p. 76)

Nowhere is intellectual beauty so deeply felt and fastidiously appreciated in its various grades and qualities as in mathematics, and only the informal appreciation of mathematical value can distinguish what is mathematics from a welter of formally similar, yet altogether trivial statements and operations.

Personal Knowledge
Chapter 6, section 10 (p. 188)

All these difficulties are but consequences of our refusal to see that mathematics cannot be defined without acknowledging its most obvious feature: namely, that it is interesting.

Personal Knowledge
Chapter 6, section 10 (p. 188)

We should declare instead candidly that we dwell on mathematics and affirm its statements for the sake of its intellectual beauty, which betokens the reality of its conceptions and the truth of its assertions. For if this passion were extinct, we would cease to understand mathematics; its conceptions would dissolve and its proofs carry no conviction. Mathematics would become pointless and would lose itself in a welter of insignificant tautologies and of Heath Robinson operations, from which it could no longer be distinguished.

Personal Knowledge
Chapter 6, section 11 (p. 192)

Modern mathematics can be kept alive only by a large number of mathematicians cultivating different parts of the same system of values: a community which can be kept coherent only by the passionate vigilance of universities, journals and meetings, fostering these values and imposing the same respect for them on all mathematicians.

Personal Knowledge
Chapter 6, section 11 (p. 192)

Pólya, George

The traditional mathematics professor of the popular legend is absentminded. He usually appears in public with a lost umbrella in each hand. He prefers to face a blackboard and to turn his back on the class. He writes *a*, he says *b*, he means *c*, but it should be *d*.

How to Solve It (p. 181)

Mathematics is being lazy. Mathematics is letting the principles do the work for you so that you do not have to do the work yourself.

Mathematics Teaching
Quoted in Marion Walter and Tom O'Brien article
Memories of George Pólya
Volume 116, September 1986 (p. 4)

Pope, Alexander
See Mystery to Mathematics fly!

The Complete Poetical Works
The Duncaid
Book IV, l. 647

Price, Bartholomew
Mathematics is the most powerful instrument, which we possess, for this purpose [to trace into their farthest results those general laws which an inductive philosophy has supplied]: in many sciences a profound knowledge of mathematics is indispensable for a successful investigation. In the most delicate researches into the theories of light, heat, and sound it is the only instrument; they have properties which no other language can express; and their argumentative processes are beyond the reach of other symbols.

Treatise on Infinitesimal Calculus
Volume 3 (p. 5)

Princeton University Catalogue (1947–1959)
Mathematics is in a sense 'the language of science,' basic to advanced work in almost all the natural sciences and in some of the social sciences.

Quoted in Eric M. Rogers
Physics for the Inquiring Mind (p. 467)

Proclus
This, therefore, is mathematics: she reminds you of the invisible form of the soul; she gives life to her own discoveries; she awakens the mind and purifies the intellect; she brings light to our intrinsic ideas; she abolishes oblivion and ignorance which are ours by birth.

Quoted in Morris Kline
Mathematical Thought from Ancient to Modern Times (p. 24)

Rayleigh, Lord
Examples . . . which might be multiplied *ad libitum*, show how difficult it often is for an experimenter to interpret his results without the aid of mathematics.

Quoted in E.T. Bell
Men of Mathematics (p. xvi)

Recorde, Robert
Beside the mathematical arts there is no infallible knowledge, except it be borrowed from them.

Quoted in Morris Kline
Mathematics and the Physical World (p. 130)

Reid, Constance
Mathematics is a world created by the mind of man, and the mathematicians are people who devote their lives to what seems to me a wonderful kind of play!

Two Year College Mathematical Journal
Quoted in G.L. Anderson
An Interview with Constance Reid
Volume 11, 1980 (p. 238)

Reid, Thomas
In mathematics it [sophistry] had no place from the beginning: Mathematicians having had the wisdom to define accurately the terms they use, and to lay down, as axioms, the first principles on which their reasoning is grounded. Accordingly we find no parties among mathematicians, and hardly any disputes.

Essays on the Intellectual Powers of Man
Essay 1, Chapter 1

The mathematician pays not the least regard either to testimony or conjecture, but deduces everything by demonstrative reasoning, from his definitions and axioms. Indeed, whatever is built upon conjecture, is improperly called science; for conjecture may beget opinion, but cannot produce knowledge.

Essays on the Intellectual Powers of Man
Essay 1, Chapter 3

Richardson, Moses
. . . I propose the following, if not as a definition, then at least as a partial description; *mathematics is persistent intellectual honesty.*

American Mathematical Monthly
Mathematics and Intellectual Honesty
Volume 59, Number 2, February 1952 (p. 73)

Rosenbaum, R.A.
Mathematical abstraction, to be considered significant, must someday pass the test of generality, of applicability, of relatedness. Mathematics too long divorced from reality, it has been said, becomes baroque, decadent, and sterile.

Mathematical Teacher
Mathematics, the Artistic Science
Volume 55, Number 7, November 1962 (p. 533)

Rózsa, Péter

The eternal lesson is that Mathematics is not something static, closed, but living and developing. Try as we may to constrain it into a closed form, it finds an outlet somewhere and escapes alive.

Playing with Infinity (p. 265)

Russell, Bertrand

Pure mathematics consists entirely of asseverations to the effect that, if such and such a proposition is true of *anything*, then such and such another proposition is true of that thing. It is essential not to discuss whether the first proposition is really true, and not to mention what the anything is, of which it is supposed to be true . . . *If* our hypothesis is about *anything*, and not about some one or more particular things, then our deductions constitute mathematics. Thus mathematics may be defined as the subject in which we never know what we are talking about, nor whether what we are saying is true.

Mysticism and Logic and Other Essays
Mathematics and the Metaphysicians (p. 75)

The true spirit of delight, the exaltation, the sense of being more than Man, which is the touchstone of the highest excellence, is to be found in mathematics as surely as in poetry.

Mysticism and Logic and Other Essays
The Study of Mathematics (p. 60)

Mathematics, rightly viewed, possess not only truth, but supreme beauty—a beauty cold and austere, like that of a sculpture . . .

Mysticism and Logic and Other Essays
The Study of Mathematics (p. 60)

. . . the rules of logic are to mathematics what those of structure are to architecture.

Mysticism and Logic and Other Essays
The Study of Mathematics (p. 61)

The discovery that all of mathematics follows inevitably from a small collection of fundamental laws is one which immeasurably enhances the intellectual beauty of the whole; to those who have been oppressed by the fragmentary and incomplete nature of most chains of deduction this discovery comes with all the overwhelming force of a revelation; like a palace emerging from the autumn mist as the traveler ascends an Italian hillside, the stately stories of the mathematical edifice appear in due order and proportion, with a new perfection in every part.

Mysticism and Logic and Other Essays
The Study of Mathematics (p. 67)

But mathematics takes us . . . into the region of absolute necessity, to which not only the actual world, but every possible world, must conform . . .

Mysticism and Logic and Other Essays
The Study of Mathematics (p. 69)

. . . mathematics is the manhood of logic . . .

Introduction to Mathematical Philosophy
Mathematics and Logic (p. 194)

In universities, mathematics is taught mainly to men who are going to teach mathematics to men who are going to teach mathematics to . . . Sometimes, it is true, there is an escape from this treadmill. Archimedes used mathematics to kill Romans, Galileo to improve the Grand Duke of Tuscany's artillery, modern physicists (grown more ambitious) to exterminate the human race. It is usually on this account that the study of mathematics is commended to the general public as worthy of State support.

Quoted in Gerald M. Weinberg
Rethinking System Analysis and Design (pp. 28–9)

I like mathematics because it is not human and has nothing particular to do with this planet or with the whole accidental universe—because like Spinoza's God, it won't love us in return.

Letter to Lady Ottoline Morrell
March 1912

Pure Mathematics is the class of all propositions of the form "p implies q," where p and q are propositions containing one or more variables, the same in the two propositions, and neither p or q contains any constants except logical constants. And logical constants are all notions definable in terms of a class of the following: Implication, the relation of a term to a class of relation, and such further notions as may be involved in the general notion of propositions of the above form. In addition to these, Mathematics *uses* a notion which is not a constituent of the propositions which it considers—namely, the notion of truth.

Principles of Mathematics (p. 1)

The nineteenth century which prides itself upon the invention of steam and evolution, might have derived a more legitimate title to fame from the discovery of pure mathematics.

International Monthly
Recent Work on the Principles of Mathematics
Volume 4, July–December 1901 (p. 83)

To create a healthy philosophy you should renounce metaphysics but be a good mathematician.

> Quoted in E.T. Bell
> *Men of Mathematics* (p. xvii)

Santayana, George

. . . mathematics is like music, freely exploring the possibilities of form. And yet, notoriously, mathematics holds true of things; hugs and permeates them far more closely than does confused and inconstant human perception; so that the dream of many exasperated critics of human error has been to assimilate all science to mathematics, so as to make knowledge safe by making it, as Locke wished, direct perception of the relations between ideas . . .

> *Realm of Truth*
> Chapter I (pp. 2–3)

If all the arts aspire to the condition of music, all the sciences aspire to the condition of mathematics.

> *Some Turns of Thought in Modern Philosophy*
> Chapter III (p. 80)

Sarton, George

Mathematics gives to science its innermost unity and cohesion, which can never be entirely replaced with props and buttresses or with roundabout connections, no matter how many of these may be introduced.

> *The Study of the History of Mathematics* (p. 4)

Sawyer, W.W.

The scientist who uses mathematics should be aware that much new mathematical knowledge is being discovered; nearly all of it will be irrelevant to his research, but he should keep his eyes open for the small piece that may be of great value to him.

> *Scientific American*
> Algebra
> Volume 211, Number 3, September 1964 (p. 78)

A point that should be borne in mind is that, generally speaking, higher mathematics is simpler than elementary mathematics. To explore a thicket on foot is a troublesome business; from an aeroplane the task is easier.

> *Prelude to Mathematics* (p. 11)

In other arts, if we see a pattern we can admire its beauty; we may feel that it has significant form, but we cannot say what the significance is. And it is much better not to try . . .

But in mathematics it is not so. In mathematics, if a pattern occurs, we can go on to ask, Why does it occur? What does it signify? And we can find answers to these questions. In fact, for every pattern that appears, a mathematician feels he ought to know why it appears.

Prelude to Mathematics (p. 23)

Any part of modern mathematics is the end-product of a long history. It has drawn on many other branches of earlier mathematics, it has extracted various essences from them and has been reformulated again and again in increasingly general and abstract forms. Thus a student may not be able to see what it is all about, in much the same way that a caveman confronted with a vitamin pill would not easily recognize it as food.

A First Look at Numerical Functional Analysis (p. 1)

Schlicter, Dean
Go down deep enough into anything and you will find mathematics.

Quoted in Margaret Joseph
The Future of Geometry
in *The Mathematics Teacher*
January 1936, Volume XXIX, Number 1 (p. 29)

Schrödinger, Erwin
. . . I do not refer to the mathematical difficulties, which eventually are always trivial, but rather to the conceptual difficulties.

Science and the Human Temperament (p. 189)

Schubert, Hermann
Whenever, therefore, a controversy arises in mathematics, the issue is not whether a thing is true or not, but whether the proof might not be conducted more simply in some other way, or whether the proposition demonstrated is sufficiently important for the advancement of the science as to deserve especial enunciation and emphasis, or finally, whether the proposition is not a special case of some other and more general truth which is just as easily discovered.

Mathematical Essays and Recreations (p. 28)

Schuttzenberger, P.
Ere long mathematics will be as useful to the chemist as the balance.

Quoted in J.W. Mellor
Higher Mathematics for Students of Chemistry and Physics (p. xvii)

Serge, Lang
I think rather that one does mathematics because one likes to do this sort of thing, and also, much more naturally, because when you have a talent for something, usually you don't have any talent for something else, and you do whatever you have talent for, if you are lucky enough to have it. I must also add that I do mathematics also because it is difficult, and it is a very beautiful challenge for the mind. I do mathematics to prove to myself that I am capable of meeting this challenge, and win it.

The Beauty of Doing Mathematics (p. 5)

Shadwell, Thomas
Wood to La. *Vaine*. 'Tis true, Madame, Sir *Positive and Poet* Ninny are excellent men, and brave Bully-Rocks; but they must grant, that neither of e'm understand Mathematicks but myself.

Sir *Posit*. Mathematicks? why, Who'se that tales of Mathematicks? Let e'm alone, let e'm alone: Now you shall see, *Stanford*.

Wood. Why, 'twas I Dear Heart.

Sir *Posit*. I Dear Heart quoth'a? I don't think you understand the principles on't; o' my Conscious you are scarce come so far yet as the squaring of the Circle, or finding out the Longitude Mathematicks . . .

The Complete Works of Thomas Shadwell
The Sullen Lovers
Act IV

Shakespeare, William
Music and poesy used to quicken you;
The mathematics and metaphysics,
Fall to them as you find your stomach serves you;
No profit grows where is no pleasure ta'en:
In brief, sir, study what you most affect.

The Taming of the Shrew
Act I, Scene 1, l. 36–40

I do present you with a man of mine,
Cunning in music and mathematics,
To instruct her fully in those sciences,
Whereof, I know, she is not ignorant.

The Taming of the Shrew
Act 2, Scene 1, l. 55–8

Shaw, George Bernard
You propound a complicated mathematical problem: give me a slate and half an hour's time, and I can produce a wrong answer.

Quoted in Evan Esar
20,000 Quips and Quotes

Shaw, James B.
Mathematics is, on the artistic side, a creation of new rhythms, orders, designs and harmonies, and on the knowledge side, is a systematic study of the various rhythms, orders, designs and harmonies. We may condense this into the statement that mathematics is, on the one side, the qualitative study of the structure of beauty, and on the other side is the creator of new artistic forms of beauty. The mathematician is at once creator and critic.

> Quoted in W.L. Schaaf (Editor)
> *Mathematics: Our Great Heritage*
> Mathematics—the Subtle Fine Art (p. 50)

. . . because mathematics contains truth, it extends its validity to the whole domain of art and the creatures of the constructive imagination. Because it contains freedom, it guarantees freedom to the whole realm of art. Because it is not primarily utilitarian, it validates the joy of imagination for the pure pleasure of imagination.

> *Lectures on the Philosophy of Mathematics* (pp. 194–5)

Shchatunovski, Samuil
It is not the job of mathematicians . . . to do correct arithmetical operations. It is the job of bank accountants.

> Quoted in George Gamow
> *My World Lines* (p. 24)

Smith, David Eugene
. . . I would rather be a dreamer without mathematics, than a mathematician without dreams.

> *The Poetry of Mathematics and Other Essays* (p. 30)

One thing that mathematics early implants, unless hindered from so doing, is the idea that here, at last, is an immortality that is seemingly tangible—the immortality of a mathematical law.

> *The Poetry of Mathematics and Other Essays* (pp. 31–2)

I know of nothing which acts as such a powerful antidote to that which I ventured to call "opinionatedness," as a study of mathematics.

> *The Poetry of Mathematics and Other Essays* (p. 35)

Smith, W.B.
Mathematics is the universal art apodictic.

> Quoted in Columbia University
> *Lectures on Science, Philosophy and Art 1907–1908* (p. 13)

Speiser, A.
. . . mathematics has liberated itself from language; and one who knows the tremendous labor put into this process and its ever-recurring surprising success, cannot help feeling that mathematics nowadays is more efficient in its particular sphere of the intellectual world than, say, the modern languages in their deplorable condition of decay or even music are on their fronts.

Klassische Stücke der Mathematik (p. 148)
Quoted in Hermann Weyl
Philosophy of Mathematics and Natural Science (p. 65)

Spencer-Brown, George
That mathematics, in common with other art forms, can lead us beyond ordinary existence, and can show us something of the structure in which all creation hangs together, is no new idea. But mathematical texts generally begin the story somewhere in the middle, leaving the reader to pick up the thread as best he can.

Laws of Form
A Note on the Mathematical Approach (p. v)

Spengler, Oswald
. . . mathematics, accessible in its full depth only to the very few, holds a quite peculiar position amongst the creation of the mind. It is a science of the most rigorous kind, like logic but more comprehensive and very much fuller; it is a true art, along with sculpture and music, as needing the guidance of inspiration and as developing under great conventions of form . . .

The Decline of the West
Volume I
Chapter II, section ii (p. 56)

And so the development of the new mathematic consists of a long, secret, and finally victorious battle against the notion of magnitude.

The Decline of the West
Volume I
Chapter II, section ix (p. 76)

Spiegel, M.R.
As I recite the alphabet to one who's only three
The world of mathematics is opened up to me.

For *a*, *b*, *c* are constants or parameters assigned
And D is a determinant or distance undefined.

The image which *e* gives to me is one I can't erase
For it can only mean for me the logarithmic base.

F, G, H are functions with appropriate domain
And *i*'s a unit vector in the Gauss or complex plane.

J's a Bessel function and another kind is K.
L's a linear operator or inductance one could say.

M and N are integers but m could be mass.
O is the number zero but $\emptyset$'s the empty class.

P and Q give odds that you will win or lose a bet.
R gives correlation of two variables you have met.

I see before me Einstein's world when I hear S and T
For they make me think of space and time and relativity.

At this point I'm so deep in thought of time and space and such
That velocity components u, v, w don't seem much.

Perhaps some day that three-year old may learn when fully grown
Why x, y, z imply for me how much there is unknown.

Mathematics Magazine
Our Mathematical Alphabet
Volume 57, Number 3, May 1984 (p. 141)

Stabler, E. Russell

Mathematics is the science of number and space. It starts from a group of self-evident truths and by infallible deduction arrives at incontestable conclusions . . . the facts of mathematics are absolute, unalterable, and eternal truths.

The Mathematics Teacher
An Interpretation and Comparison of Three Schools of
Thought in the Foundations of Mathematics
Volume 26, January 1935 (p. 6)

Steen, Lynn

. . . despite an objectivity about mathematical results that has no parallel in the world of art, the motivation and standards of creative mathematics are more like those of art than like those of science. Aesthetic judgments transcend both logic and applicability in the ranking of mathematical theorems: beauty and elegance have more to do with the value of a mathematical idea than does either strict truth or possible utility.

Mathematics Today: Twelve Informal Essays
Mathematics Today (p. 10)

Steinmetz, Charles Proteus

Mathematics is the most exact science, and its conclusions are capable of absolute proof. But this is so only because mathematics does not *attempt* to draw absolute conclusions. All mathematical truths are relative, conditional.

Quoted in E.T. Bell
Men of Mathematics (p. xvii)

Stendhal [Henri Beyle]
I used to love mathematics for its own sake, and I still do, because it allows for no hypocrisy and no vagueness . . .

The Life of Henri Brulard
Chapter 10

Sterne, Laurence
[My Uncle Tody] proceeded next to Galileo and Torricellius, wherein, by certain Geometrical rules, infallibly laid down, he found the precise part to be a "Parabola"—or else an "Hyperbola,"—and that the parameter, or "latus rectum," of the conic section of the said path, was to the quantity and amplitude in a direct ratio, as the whole line to the sine of double the angle of incidence, formed by the breech upon an horizontal line;—and that the semiparameter,—stop! my dear uncle Toby—stop!

Quoted in James R. Newman
The World of Mathematics
Volume II (p. 734)

Stewart, Dugald
. . . the study of it [mathematics] is peculiarly calculated to strengthen the power of steady and concatenated thinking—a power which, in all the pursuits of life, whether speculative or active, is one of the most valuable endowments we can possess. This command of attention, however, it may be proper to add, is to be acquired, not by the practice of modern methods, but by the study of the Greek geometry . . .

The Collected Works of Dugald Stewart
Volume IV (p. 201)

Stewart, Ian
Not all ideas are mathematics; but all good mathematics must contain an idea.

The Nature of Mathematics (p. 6)

Mathematics is much like the Mississippi. There are sideshoots and dead ends and minor tributaries; but the mainstream is there, and you can find it where the current—the mathematical power—is strongest. Its delta is research mathematics: it is growing, it is going somewhere (but it may not always be apparent where), and what today looks like a major channel may tomorrow clog up with silt and be abandoned. Meanwhile a minor trickle may suddenly open out into a roaring torrent. The best mathematics always enriches the mainstream, sometimes by diverting it in an entirely new direction.

The Nature of Mathematics (p. 12)

Mathematics is to nature as Sherlock Holmes is to evidence.

Nature's Numbers (p. 2)

Mathematics is not just a collection of isolated facts: it is more like a landscape; it has an inherent geography that its users and creators employ to navigate through what would otherwise be an impenetrable jungle.

Nature's Numbers (p. 38)

Stone, Marshall H.

. . . science is reasoning; reasoning is mathematics; and, therefore, science is mathematics.

Bulletin of the American Mathematical Association
Mathematics and the Future of Science
Volume 63, Number 2, March 1957 (p. 61)

Struik, Dirk J.

Mathematics is a vast adventure of ideas; its history reflects some of the noblest thoughts of countless generations.

A Concise History of Mathematics
Introduction (p. xi)

Sullivan, J.W.N.

Mathematics is the expression of this life so far as the intellect is concerned.

Aspects of Science (p. 193)

. . . a history of mathematics is largely a history of discoveries which no longer exist as separate items, but are merged into some more modern generalization, these discoveries have not been forgotten or made valueless. They are not dead, but transmuted.

The History of Mathematics in Europe
Introduction (p. 10)

Sylvester, James Joseph

May not Music be described as the Mathematic of sense, Mathematic as the Music of the reason? the soul of each the same! Thus the musician *feels* Mathematic, the mathematician *thinks* Music,—Music the dream, Mathematic the working life—each to receive its consummation from the other when the human intelligence, elevated to its perfect type, shall shine forth glorified in some future Mozart–Dirichlet or Beethoven–Gauss—a union already not indistinctly foreshadowed in the genius and labours of a Helmholtz!

The Collected Mathematical Papers of James Joseph Sylvester
Volume II
On Newton's Rule for the Discovery of Imaginary Roots
(p. 419, fn)

The world of ideas which it [mathematics] discloses or illuminates, the contemplation of divine beauty and order which it induces, the harmonious connection of its parts, the infinite hierarchy and absolute evidence of the truths with which mathematical science is concerned, these, and such like, are the surest grounds of its title of human regard, and would remain unimpaired were the plan of the universe unrolled like a map at our feet, and the mind of man qualified to take in the whole scheme of creation at a glance.

The Collected Mathematical Papers of James Joseph Sylvester
Volume II
Presidential Address to Section 'A' of the British Association (p. 659)

. . . I think it would be desirable that this form of the word [mathematics] should be reserved for the application of the science, and that we should use mathematic in the singular to denote the science itself, in the same was as we speak of logic, rhetoric, or (own sister to algebra) music.

The Collected Mathematical Papers of James Joseph Sylvester
Volume II
Presidential Address to Section 'A' of the British Association (p. 659)

The object of pure Physic is the unfolding of the laws of the intelligible world; the object of pure Mathematic that of the unfolding the laws of human intelligence.

The Collected Mathematical Papers of James Joseph Sylvester
Volume III
On a Theorem Connected with Newton's Rule (p. 424)

Number, place, and combinations . . . the three intersecting but distinct spheres of thought to which all mathematical ideas admit of being referred.

Philosophical Magazine (1844) (p. 285)

Mathematics is not a book confined within a cover and bound between brazen clasps, whose contents it needs only patience to ransack; it is not a mine, whose treasures may take long to reduce into possession, but which fill only a limited number of veins and lodes; it is not a soil, whose fertility can be exhausted by the yield of successive harvests; it is not a continent or an ocean, whose area can be mapped out and its contour defined; it is as limitless as that space which it finds too narrow for its aspirations; its possibilities are as infinite as the worlds which are forever crowding in and multiplying upon the astronomer's gaze; it is as incapable of being restricted within assigned boundaries or being reduced to definitions of permanent validity, as the consciousness of life.

Quoted in John D. Barrow
Pi in the Sky (p. 124)

Synge, J.L.
Logic is the railway track along which the mind glides easily. It is the axioms that determine our destination by setting us on this track or the other, and it is in the matter of choice of axioms that applied mathematics differs most fundamentally from pure. Pure mathematics is controlled (or should we say "uncontrolled"?) by a principle of ideological isotropy: any line of thought is as good as another, provided that it is logically smooth. Applied mathematics on the other hand follows only those tracks which offer a view of natural scenery; if sometimes the track dives into a tunnel it is because there is prospect of scenery at the far end.

American Mathematical Monthly
Postcards on Applied Mathematics
Volume 46, Number 3, March 1939 (p. 156)

Szegö, Gábor
Mathematics is a human activity almost as diverse as the human mind itself.

Quoted in Jozef Kürschak
Hungarian Problem Book
Volume 1 (p. 6)

Teller, Edward
Science attempts to find logic and simplicity in nature. Mathematics attempts to establish order and simplicity in human thought.

The Pursuit of Simplicity (p. 17)

Thompson, William Robin
In short, mathematics is not a substitute for experiment: neither is mathematics experimental.

Science and Common Sense
The Use and Abuse of Mathematics (p. 124)

Thomson, William [Lord Kelvin]
A single curve, drawn in the manner of the curve of prices of cotton describes all that the ear can possibly hear as a result of the most complicated musical performances . . . That to my mind is a wonderful proof of the potency of mathematics.

Quoted in E.T. Bell
Men of Mathematics (p. xvi)

Mathematics is the only good metaphysics.

Quoted in E.T. Bell
Men of Mathematics (p. xvii)

Mathematics is the only true metaphysics.

In S.P. Thompson
The Life of William Thomson Baron Kelvin of Largs
Volume II
Views and Opinions (p. 1139)

Thoreau, Henry David

We have heard much about the poetry of mathematics, but very little of it has yet been sung. The ancients had a juster notion of their poetic value than we. The most distinct and beautiful statements of any truth must take at last the mathematical form.

A Week on the Concord and Merrimac Rivers
Friday (p. 323)

Thurston, William

I think mathematics is a vast territory. The outskirts of mathematics are the outskirts of mathematical civilization. There are certain subjects that people learn about and gather together. Then there is a sort of inevitable development in those fields. You get to the point where a certain theorem is bound to be proved, independent of any particular individual, because it is just in the path of development.

Quoted in D. Albers, G. Alexanderson, and C. Reid (Editors)
More Mathematical People
William P. Thurston (p. 332)

Tolstoy, Leo

"Mathematics are most important, madam! I don't want to have you like our silly ladies. Get used to it and you'll like it," and he patted her cheek. "It will drive all the nonsense out of your head."

War and Peace
Book I, Chapter 25

A modern branch of mathematics having acquired the art of dealing with the infinitely small can now yield solutions in other more complex problems of motion which used to appear insoluble.

The modern branch of mathematics, unknown to the ancients, when dealing with problems of motion admits the conception of the infinitely small, and so conforms to the chief condition of motion (absolute continuity) and thereby corrects the inevitable error which the human mind cannot avoid when it deals with separate elements of motion instead of examining continuous motion.

War and Peace
Book XI, Chapter 1

Traditional teaching of mathematics
Ignore the facts.

Traditional teaching of physics
Stick to the facts.

Trevelyan, George Otto

He gave himself diligently to mathematics, which he liked "vastly". "I believe they are useful," he writes, "and I am sure they are entertaining, which is alone enough to recommend them to me."

> *The Early History of Charles James Fox*
> Chapter II (p. 50)

Trudeau, Richard J.

Pure mathematics is the world's best game. It is more absorbing than chess, more of a gamble than poker, and lasts longer than Monopoly. It's free. It can be played anywhere—Archimedes did it in a bathtub. It is dramatic, challenging, endless, and full of surprises.

> *Dots and Lines* (p. 9)

Twain, Mark

We could use up two Eternities in learning all that is to be learned about our own world and the thousands of nations that have arisen and flourished and vanished from it. Mathematics alone would occupy me eight million years.

> *Notebook*
> Chapter 22 (p. 170)

Ulam, Stanislaw

Do not lose your faith. A mighty fortress is our mathematics. Mathematics will rise to the challenge, as it always has.

> Quoted in Heinz R. Pagels
> *The Dreams of Reason* (p. 94)

Unknown

If there be a pedagogical devil and hell, a Duncaid kingdom of darkness and stupidity ever insidiously warring upon the spirit of sweetness, light, and truth which is the holy spirit of teaching, they were never more strongly entrenched than in elementary mathematics. Here about every pedagogical disease ever known has flourished. This has been the sickest of all sick topics, immune to no epidemic or contagion and more diversely doctored and prescribed for by every kind of therapy.

> Quoted in G. Stanley Hall
> *Educational Problems*
> Volume II
> Chapter XVIII (p. 342)

Once upon a time $(1/t)$ pretty little Polly Nomial was strolling across a field of vectors when she came to the boundary of a singularly large matrix. Now Polly was convergent, and her mother had made it an absolute condition that she must never enter such an array without her brackets on. Polly, however, who had changed her variables that morning and was feeling particularly badly behaved, ignored this condition on the basis that it was insufficient and made her way in amongst the complex elements. Rows and columns closed in on her from all sides. Tangents approached her surface. She became tensor and tensor. Quite suddenly two branches of a hyperbola touched her at a single point. She oscillated violently, lost all sense of directrix, and went completely divergent. She tripped over a square root that was protruding from the erf and plunged headlong down a steep gradient. When she rounded off once more, she found herself inverted, apparently alone, in a non-Euclidean space.

She was being watched, however. That smooth operator, Curly Pi, was lurking inner product. As his eyes devoured her curvilinear coordinates, a singular expression crossed his face. He wondered, "Was she still convergent?" He decided to integrate properly at once.

Hearing a common fraction behind her, Polly rotated and saw Curly Pi approaching with his power series extrapolated. She could see at once by his degenerate conic and dissipative that he was bent on no good.

"Arcsinh," she gasped.

"Ho, ho," he said, "What a symmetric little asymptote you have. I can see your angles have lots of secs."

"Oh sir," she protested, "keep away from me I haven't got my brackets on."

"Calm yourself, my dear," said our suave operator, "your fears are purely imaginary."

"I, I," she thought, "perhaps he's not normal but homologous."

"What order are you?" the brute demanded.

"Seventeen," replied Polly.

Curly leered "I suppose you've never been operated on."

"Of course not," Polly replied quite properly, "I'm absolutely convergent."

"Come, come," said Curly, "let's off to a decimal place I know and I'll take you to the limit."

"Never," gasped Polly.

"Abscissa," he swore, using the vilest oath he knew. His patience was gone. Coshing her over the coefficient with a log until she was powerless, Curly removed her discontinuities. He stared at her significant places, and began smoothing out her points of inflection. Poor Polly. The algorithmic method was now her only hope. She felt his digits tending to her asymptotic limit. Her convergence would soon be gone forever.

There was no mercy, for Curly was a Heaviside operator. Curly's radius squared itself; Polly's loci quivered. He integrated by parts. He integrated by partial fractions. After he cofactored, he performed Runge–Kutta on her. The complex beast even went all the way around and did a contour integration. What an indignity—to be multiply connected on her first integration. Curly went on operating until he completely satisfied her hypothesis, then he exponentiated and became completely orthogonal.

When Polly got home that night, her mother noticed that she was no longer piecewise continuous, but had been truncated in several places. But it was to late to differentiate now. As the months went by, Polly's denominator increased monotonically. Finally she went to L'Hôpital and generated a small but pathological function which left surds all over the place and drove Polly to deviation.

The moral of our sad story is this: "If you want to keep your expressions convergent, never allow them a single degree of freedom."

Source unknown

A mathematician named Klein
Thought the Moebius band was divine.
  Said he, "If you glue
    The edges of two,
You'll get a weird bottle like mine.

Quoted in Arnold O. Allen
Probability, Statistics, and Queueing Theory with
Computer Science Applications (p. 436)

Math was always my worst subject.

Source unknown

You can't argue with mathematics.

From the movie The Unsuspected

In mathematics, fractions speak louder than words.

Source unknown

There was a young man of Nepal
Who had a mathematical ball;
  $(W^3 \times \pi) - 8 = 4/3\sqrt{0}$
  (The Cube of its weight
    Times Pi, minus eight
Is four thirds of the root of fuck all)

Source unknown

Mathematics is music for the mind;
Music is mathematics for the soul.

Quoted in Stanley Gudder
A Mathematical Journey (p. 58)

There was a professor of math
Who was thrilling his girl in her bath.
　The soap slipped from reach
　And plugged up her breech,
So he finished the job in her ath.

In G. Legman (Editor)
The New Limerick
1011

Evolution of test questions in mathematics

1960's Arithmetic Test: "A logger cuts and sells a truckload of lumber for $100. His cost of production is four-fifths of that amount. What is his profit?"

1970's New-Math Test: "A logger exchanges a set {L} of lumber for a set {M} of money. The cardinality of set M is 100. The set C of production costs contains 20 fewer points. What is the cardinality of set P of profits?"

1980's "Dumbed down" version: " A logger cuts and sells a truckload of lumber of $100. His cost is $80, his profit is $20. Find and circle the number 20."

1990's version: "An unenlightened logger cuts down a beautiful stand of 100 trees in order to make a $20 profit. Write an essay explaining how you feel about this as a way to make money. Topic for discussion: How did the forest birds and squirrels feel?"

Source unknown

The judge, Pope, and a mathematics professor were arguing about who was the most powerful of the three. The judge said, "everytime I walk into court, everyone must rise to salute me" and the Pope rebuked, "Oh that's nothing. People kneel in front of me just to kiss my ring." The math professor, a little confused, timidly commented, "Well, I don't know, but every time I walk in class, all the students lower their heads and mumble, 'Oh God!'".

Source unknown

The biologist thinks
He is a chemist,
The chemist thinks
He is a physicist,
The physicist thinks

He is a God,
And God thinks
She is a mathematician.

Source unknown

Viereck, G.S.
Eldridge, Paul
We are two parallel lines drawn very close to each other . . . So close indeed that no third line, however thin, could be drawn between them.

Will the two parallel lines ever meet?

Yes, In infinity.

Ali Hasan! I exclaimed, had you ever dreamed that there was so much poetry and pathos and sorrow in mathematics?

My First Two Thousand Years
Chapter XLIV (p. 218)

Voltaire
When we cannot use the compass of mathematics or the torch of experience . . . it is certain that we cannot take a single step forward.

Quoted in David M. Burton
The History of Mathematics (p. 232)

All certainty which does not consist in mathematical demonstration is nothing more than the highest probability; there is no other historical certainty.

The Portable Voltaire
Philosophical Dictionary
Miscellany (p. 223)

von Neumann, John
. . . in mathematics you don't understand things. You just get used to them.

Quoted in G. Zukav
The Dancing Wu Li Masters
(p. 226, fn)

. . . mathematics is not an empirical science, or at least that it is practiced in a manner which differs in several decisive respects from the techniques of the empirical sciences.

Collected Works
Volume I
The Works of the Mind
The Mathematician (p. 6)

Mathematics falls into a great number of subdivisions, differing from one another widely in character, style, aims, and influence. It shows the very opposite of the extreme concentration of theoretical physics. A good theoretical physicist may today still have a working knowledge of more than half of his subject. I doubt that any mathematician now living has much of a relationship to more than a quarter.

Collected Works
Volume I
The Works of the Mind
The Mathematician (p. 6)

Waismann, Friedrich

We could compare mathematics so formalized to a game of chess in which the symbols correspond to the chessmen; the formulae, to definite positions of the men on the board; the axioms, to the initial positions of the chessmen; the directions for drawing conclusions, to the rules of movement; a proof, to a series of moves which leads from the initial position to a definite configuration of the men.

Introduction to Mathematical Thinking
Chapter 6 (pp. 76–7)

Wall, H.S.

Mathematics is a creation of the mind. To begin with, there is a collection of things, which exist only in the mind, assumed to be distinguishable from one another; and there is a collection of statements about these things, which are taken for granted. Starting with the assumed statements concerning these invented or imagined things, the mathematician discovers other statements, called theorems, and proves them as necessary consequences. This, in brief, is the pattern of mathematics. The mathematician is an artist whose medium is the mind and whose creations are ideas.

Creative Mathematics (p. 3)

Walton, Izaak

For Angling may be said to be so like the Mathematicks, that it can never be fully learnt . . .

The Compleat Angler
To All Readers (p. xli)

Weaver, Warren

There is a common tendency to consider mathematics so strange, subtle, rigorous, difficult and deep a subject that if a person is a mathematician he is of course a "great mathematician"—there being, so to speak, no small giants. This is very complimentary, but unfortunately not necessarily true.

Scientific American
Lewis Carroll: Mathematician
Volume 194, Number 4, April 1956

Webster, John

Bosola: Didst thou never study mathematics?

Old Lady: What's that, sir?

Bosola: Why, to know the trick how to make many lines meet in one centre.

Duchess of Malfi
Act II, scene ii

Weil, André

God exists since mathematics is consistent, and the Devil exists since we cannot prove it.

Quoted in John D. Barrow
The World within the World (p. 254)

Mathematics has this peculiarity, that it is not understood by non-mathematicians.

Oeuvres Scientifiques
Organisation et désorganisation en mathématique
Volume II (p. 465)

Wells, H.G.

. . . the new mathematics is a sort of supplement to language, affording a means of thought about form and quantity and a means of expression, more exact, compact, and ready than ordinary language. The great body of physical science, a great deal of the essential facts of financial science, and endless social and political problems are only accessible and only thinkable to those who have had a sound training in mathematical analysis, and the time may not be very remote when it will be understood that for complete initiation as an efficient citizen of one of the new great complex world wide states that are now developing, it is as necessary to be able to compute, to think in averages and maxima and minima, as it is now to be able to read and to write.

Mankind in the Making
Chapter VI (p. 204)

Westaway, F.W.

Mathematics, like all other subjects, has now to take its turn under the microscope and reveal to the world any weaknesses there may be in its foundations.

Quoted in E.T. Bell
Men of Mathematics (p. 555)

Weyl, Hermann

No Hilbert will be able to assure us of consistency forever; we must be content if a simple axiomatic system of mathematics has met the test of our elaborate mathematical experiments so far . . . A truly realistic mathematics should be conceived, in line with physics, as a branch of the theoretical construction of the one real world, and should adopt the same sober and cautious attitude toward hypothetic extensions of its foundations as is exhibited by physics.

Philosophy of Mathematics and Natural Science (p. 235)

In these days the angel of topology and the devil of abstract algebra fight for the soul of each individual mathematical domain.

Quoted in Morris Kline
Mathematical Thought from Ancient to Modern Times (p. 924)

Kierkegaard once said religion deals with what concerns man unconditionally. In contrast (but with equal exaggeration) one may say that mathematics talks about the things which are of no concern at all to man. Mathematics has the inhuman quality of starlight, brilliant and sharp, but cold. But it seems an irony of creation that man's mind knows how to handle things the better the farther removed they are from the center of his existence. Thus we are cleverest where knowledge matters least: in mathematics, especially in number theory.

American Mathematical Monthly
A Half-century of Mathematics
Volume 58, October 1951 (p. 523)

Without the concepts, methods and results found and developed by previous generations right down to Greek antiquity one cannot understand either the aims or the achievements of mathematics in the last fifty years.

Quoted in Morris Kline
Mathematical Thought from Ancient to Modern Times (p. 101)

Mathematics is the science of the infinite, its goal the symbolic comprehension of the infinite with human, that is finite, means.

The Open World
Chapter I (p. 7)

In mathematics the inquiry into the genuineness or non-genuineness of the inner working of our entire western culture urges towards a more rigorous decision than can be attained in the other hazier fields of knowledge.

The Rice Institute Pamphlet
Gravitation and the Electron
Mathematical Lectures (p. 246)
Volume 16, Number 4

. . . it is the function of mathematics to be at the service of the natural sciences.

Philosophy of Mathematics and Natural Science (p. 61)

The problems of mathematics are not isolated problems in a vacuum; there pulses in them the life of ideas which realize themselves *in concerto* through our human endeavors in our historical existence, but forming an indissoluble whole transcend any particular science.

> Quoted in K. Chandrasekharan
> *Hermann Weyl* (p. 84)

Whetham, W.C.D.
. . . mathematics is but the higher development of symbolic logic.

> *The Recent Development of Physical Science* (p. 34)

White, William Frank
Mathematics, the science of the ideal, becomes the means of investigating, understanding and making known the world of the real. The complex is expressed in terms of the simple. From one point of view mathematics may be defined as the science of successive substitutions of simpler concepts for more complex . . .

> *A Scrap-book of Elementary Mathematics* (p. 215)

He must be a "practical" man who can see no poetry in mathematics.

> *A Scrap-book of Elementary Mathematics* (p. 208)

Mathematics is the science of definiteness, the necessary vocabulary of those who know.

> *A Scrap-book of Elementary Mathematics* (p. 7)

Whitehead, Alfred North
Mathematics in its widest significance is the development of all types of formal, necessary, deductive reasoning.

> *A Treatise on Universal Algebra*
> Preface (p. vi)

The ideal of mathematics should be to erect a calculus to facilitate reasoning in connection with every province of thought, or of external experience, in which the succession of thoughts, or of external experience, in which the succession of thoughts, or of events can be definitely ascertained and precisely stated. So that all serious thought which is not philosophy, or inductive reasoning, or imaginative literature, shall be mathematics developed by means of a calculus.

> *A Treatise on Universal Algebra*
> Preface (p. viii)

The whole of Mathematics consists in the organization of a series of aids to the imagination in the process of reasoning.

> *A Treatise on Universal Algebra*
> Chapter I (p. 12)

. . . the pursuit of mathematics is a divine madness of the human spirit, a refuge from the goading urgency of contingent happenings . . .

> Quoted in Steve J. Heims
> *From Mathematics to the Technologies of Life and Death* (p. 116)

All mathematical calculations about the course of nature must start from some assumed law of nature . . . Accordingly, however accurately we have calculated that some event must occur, the doubt always remains— Is it true?

> *An Introduction to Mathematics*
> Chapter 3 (p. 16)

Philosophers, when they have possessed a thorough knowledge of mathematics, have been among those who have enriched the science with some of its best ideas. On the other hand it must be said that, with hardly an exception, all the remarks on mathematics made by those philosophers who have possessed but a slight or hasty or late-acquired knowledge of it are entirely worthless, being either trivial or wrong.

> *An Introduction to Mathematics*
> Chapter 9 (pp. 81–2)

The anxious precision of modern mathematics is necessary for accuracy, . . . it is necessary for research. It makes for clearness of thought and for fertility in trying new combinations of ideas. When the initial statements are vague and slipshod, at every subsequent stage of thought, common sense has to step in to limit applications and to explain meanings. Now in creative thought common sense is a bad master. Its sole criterion for judgment is that the new ideas shall look like the old ones, in other words it can only act by suppressing originality.

> *Introduction to Mathematical Science* (p. 157)

It is a safe rule to apply that, when a mathematical or philosophical author writes with a misty profundity, he is talking nonsense.

> *Introduction to Mathematical Science*
> Chapter 15

There is no more common error than to assume that, because prolonged and accurate mathematical calculations have been made, the application of the result to some fact of nature is absolutely certain.

> *Introduction to Mathematical Science*
> Chapter 3

The essence of applied mathematics is to know what to ignore.

> Quoted in R.A. Fisher's Presidential Address
> *Sankhyah* (p. 16)

The science of Pure Mathematics, in its modern developments, may claim to be the most original creation of the human spirit.

Science and the Modern World
Mathematics as an Element in the History of Thought (p. 29)

For mathematics is the science of the most complete abstractions to which the human mind can attain.

Science and the Modern World
Mathematics as an Element in the History of Thought (p. 51)

I will not go so far as to say that to construct a history of thought without profound study of the mathematical ideas of successive epochs is like omitting Hamlet from the play which is named after him. That would be claiming too much. But it is certainly analogous to cutting out the part of Ophelia. This simile is singularly exact. For Ophelia is quite essential to the play, she is very charming,—and a little mad. Let us grant that the pursuit of mathematics is a divine madness of the human spirit, a refuge from the goading urgency of contingent happenings.

Science and the Modern World
Mathematics as an Element in the History of Thought (p. 31)

The study of mathematics is apt to commence in disappointment. The important applications of the science, the theoretical interest of its ideas, and the logical rigour of its methods, all generate the expectation of a speedy introduction to processes of interest. We are told that by its aid the stars are weighed and the billions of molecules in a drop of water are counted. Yet, like the ghost of Hamlet's father, this great science eludes the efforts of our mental weapons to grasp it—"'Tis here, 'tis there, 'tis gone"—and what we do see does not suggest the same excuse for illusiveness as sufficed for the ghost, that it is too noble for our gross methods.

An Introduction To Mathematics
Chapter 1 (p. 1)

The first acquaintance which most people have with mathematics is through arithmetic . . . Arithmetic, therefore, will be a good subject to consider in order to discover, if possible, the most obvious characteristic of the science. Now, the first noticeable fact about arithmetic is that it applies to everything, to tastes and to sounds, to apples and to angels, to the ideas of the mind and to the bones of the body.

An Introduction to Mathematics
Chapter 1 (p. 2)

Wiener, Norbert
Mathematics is too arduous and uninviting a field to appeal to those to whom it does not give great rewards. These rewards are of

exactly the same character as those of the artist. To see a difficult, uncompromising material take living shape and meaning is to be Pygmalion, whether the material is stone or hard, stonelike logic. To see meaning and understanding come where there has been no meaning and no understanding is to share the work of a demiurge. No amount of technical correctness and no amount of labor can replace this creative moment, whether in the life of a mathematician or in that of a painter or musician. Bound up with it is a judgment of values, quite parallel to the judgment of values that belongs to the painter or the musician. Neither the artist nor the mathematician may be able to tell you what constitutes the difference between a significant piece of work and an inflated trifle; but if he is never able to recognize this in his own heart, he is no artist and no mathematician.

Ex-Prodigy (p. 212)

Wigner, Eugene
The enormous usefulness of mathematics in the natural sciences is something bordering on the mysterious.

Communications in Pure and Applied Mathematics
The Unreasonable Effectiveness of Mathematics in the Natural Sciences
Volume 13, February 1960 (p. 2)

. . . mathematics is the science of skillful operations with concepts and rules invented just for this purpose.

Communications in Pure and Applied Mathematics
The Unreasonable Effectiveness of Mathematics in the Natural Sciences
Volume 13, February 1960 (p. 2)

The miracle of the appropriateness of the language of mathematics for the formulation of the laws of physics is a wonderful gift, which we neither understand nor deserve.

Communications in Pure and Applied Mathematics
The Unreasonable Effectiveness of Mathematics in the Natural Sciences
Volume 13, February 1960 (p. 14)

Willerding, Margaret F.
It is strange but true that most of the greatest strides in mathematics were made at a time and in an atmosphere when the need for mathematics was the least. Mathematics flourishes when it is free to follow any course it desires and when there is no pressure for practical results limiting its scope and freedom.

School Science and Mathematics
The Uselessness of Mathematics
Part II
Volume LXVIII, Number 6, June 1968 (p. 495)

Winsor, Frederick
Three jolly sailors from Blaydon-on-Tyne
They went to sea in a bottle by Klein.
Since the sea was entirely inside the hull
The scenery seen was exceedingly dull.

The Space Child's Mother Goose

Wittgenstein, Ludwig
A mathematical proof must be perspicuous.

Remarks on the Foundations of Mathematics
Appendix II, 1 (p. 65e)

. . . mathematics is a MOTLEY of techniques and proofs.

Remarks on the Foundations of Mathematics
Appendix II, 46 (p. 84e)

There is no religious denomination in which the misuse of metaphysical expressions has been responsible for so much sin as it has in mathematics.

Culture and Value (p. 1e)

With my full philosophical rucksack I can only climb slowly up the mountain of mathematics.

Culture and Value (p. 2e)

Wright, Frank Lloyd
You can study mathematics all your life and never do a bit of thinking.

Quoted in Robert B. Heywood (Editor)
The Works of the Mind
The Architect (pp. 57–8)

Zee, Anthony
Toward the end of the last century, many physicists felt that the mathematical description of physics was getting ever more complicated. Instead, the mathematics involved has become ever more abstract, rather than more complicated. The mind of God appears to be abstract but not complicated. He also appears to like group theory.

Fearful Symmetry
Chapter 9 (p. 132)

Zelazny, Roger
An ellipse is fine for as far as it goes,
But modesty, away!
If I'm going to see Beauty without her clothes
Give me hyperbolas any old day.

Quoted in William L. Burke
Spacetime, Geometry, Cosmology (p. 22)

MAXIMUM AND MINIMUM

Euler, Leonhard
For since the fabric of the Universe is most perfect and the work of a most wise creator, nothing at all takes place in the Universe in which some rule of maximum or minimum does not appear.

Quoted in Jerrold E. Marsden and Anthony J. Tromba
Vector Calculus (p. 199)

METHOD

Egler, Frank E.
Concepts are games we play with our heads; methods are games we play with our hands, which at times are so handy they can be played without a head.

The Way of Science (p. 34)

For all its value, the application of a method, alone, is not science, any more than a pile of bricks is architecture. I would sooner trust a good mind without a method than a good method without a mind.

The Way of Science (p. 36)

Hilbert, David
. . . for he who seeks for methods without having a definite problem in mind seeks for the most part in vain.

Bulletin of the American Mathematical Society
Hilbert: Mathematical Problems
Volume 8, July 1902 (p. 444)

Pólya, George
My method to overcome a difficulty is to go round it.

How to Solve It (p. 181)

What is the difference between a method and device? A method is a device which you use twice.

How to Solve It (p. 181)

Serge, Corrado
Many times a scientific truth is placed as it were on a lofty peak, and to reach it we have at our disposal at first only dark paths along perilous slopes whence it is easy to fall into the abysses where dwells error; only after we have reached the peak by these paths is it possible to lay out safe roads which lead there without peril. Thus it has frequently happened

that the first way of obtaining a result has not been quite satisfactory, and that only *afterwards* did the science succeed in completing the demonstration. Certainly also a mathematician can not be really content with a result which he has obtained by non-rigorous methods; he will not feel sure of it until he has rigorously proved it. But he will not reject summarily these imperfect methods in the case of difficult problems when he is unable to substitute better ones, since the history of the science precisely shows what service such methods have always rendered.

Bulletin of the American Mathematical Society
2nd Series
On Some Tendencies in Geometric Investigations
Volume 10, June 1904 (pp. 453–4)

My method to overcome a difficulty is to go round it.
George Pólya – (See p. 259)

MISTAKE

Gombrich, E.H.

In order to learn, we must make mistakes, and the most fruitful mistakes which nature could have implanted in us would be the assumption of even greater simplicities than we are likely to meet in this bewildering world of ours . . . To probe a hole we first use a straight stick to see how far it takes us. To probe the visible world we use the assumption that things are simple until they prove to be otherwise.

Quoted in John Pottage
Geometrical Investigations (p. 15)

Huxley, Thomas H.

. . . there is the greatest practical benefit in making a few failures early in life.

Science and Education
On Medical Education (p. 306)

Siegel, Eli

If a mistake is not a stepping stone, it is a mistake.

Damned Welcome
Aesthetic Realism Maxims
Part I, #139 (p. 39)

MODEL

Eigen, Manfred
A theory has only the alternative of being right or wrong. A model has a third possibility: it may be right, but irrelevant.

The Physicist's Conception of Nature
Edited by Jagdish Mehra (p. 618)

Kaplan, Abraham
The words "model" and "mode" have, indeed, the same root; today, model building is science *à la mode*.

The Conduct of Inquiry
Chapter VII, Section 30 (p. 258)

Karlin, Samuel
The purpose of models is not to fit the data but to sharpen the questions.

11th R.A. Fisher Memorial Lecture
Royal Society 20 April 1983

Jeans, Sir James Hopwood
The making of models or pictures to explain mathematical formulae and the phenomena they describe, is not a step towards, but a step away from, reality; it is like making graven images of a spirit.

The Mysterious Universe
Into Deep Waters (p. 176)

Robertson, Sir Dennis
As soon as I could safely toddle
My parents handed me a Model.
My brisk and energetic pater
Provided the accelerator.
My mother, with her kindly gumption,
The function guiding my consumption;
And every week I had from her

A lovely new parameter,
With lots of little leads and lags
In pretty parabolic bags.

With optimistic expectations
I started on my explorations,
And swore to move without a swerve
Along my sinusoidal curve.
Alas! I knew how it would end:
I've mixed the cycle and the trend,
And fear that, growing daily skinnier,
I have at length become non-linear.
I wander glumly round the house
As though I were exogenous,
And hardly capable of feeling
The difference 'tween floor and ceiling.
I scarcely now, a pallid ghost,
Can tell *ex-ante* from *ex-post*:
My thoughts are sadly inelastic,
My acts incurably stochastic.

The Non-Econometrician's Lament
Source unknown

Thomson, William [Lord Kelvin]
I never satisfy myself until I can make a mechanical model of a thing. If I can make a mechanical model, I understand it.
Baltimore Lectures on Molecular Dynamics, and the Wave Theory of Light (p. 270)

Unknown
The sciences do not try to explain, they hardly even try to interpret, they mainly make models.
Source unknown

von Neumann, John
. . . the sciences do not try to explain, they hardly even try to interpret, they mainly make models.
Collected Works
Volume VI
Method in the Physical Sciences (p. 492)

MOEBIUS STRIP

Schenck, Hilbert Jr.
The topologist's mind came unguided
When his theories, some colleagues derided.
  Out of Moebius strips
  Paper dolls he now snips,
Non-Euclidean, closed, and one-sided.

The Magazine of Fantasy and Science Fiction
Snip, Snip (p. 86)
Volume 17, Number 3, September 1959

Unknown
A mathematician confided
That a Moebius band is one-sided.
  And you'll get quite a laugh
  If you cut one in half,
For it stays in one piece when divided.

Quoted in Clifton Fadiman
The Mathematical Magpie (p. 292)

A mathematician named Klein
Thought the Moebius band was divine.
  Said he, "If you glue
  The edges of two,
You'll get a weird bottle like mine."

Quoted in Clifton Fadiman
The Mathematical Magpie (p. 293)

Narcissus achieved his ambition:
He was taught by a mathematician
  To perform with great ease
  A moebius-strip tease,
With an auto-erotic emission.

In G. Legman (Editor)
The New Limerick
2242

Yes, he screwed her, but under great tension:
'Twas done with severe apprehension.
  She possessed (to be blunt)
  A true Klein-bottle cunt—
Now his prick's in another dimension.

In G. Legman (Editor)
The New Limerick
2647

Winsor, Frederick
Flappity, Floppity, Flip!
The Mouse on the Moebius Strip.
  The Strip resolved,
  The Mouse dissolved
In a chronodimensional skip.

The Space Child's Mother Goose
14

An intellectual is a highly educated man who can't do
arithmetic with his shoes on, and is proud of his lack.
Robert A. Heinlein – (See p. 14)

MULTIPLICATION

Carroll, Lewis
Four times five is twenty, and four times six is thirteen, and four times seven is—oh dear! I shall never get to twenty at that rate!

The Complete Works of Lewis Carroll
Alice's Adventures in Wonderland
The Pool of Tears

Chekhov, Anton
There is no national science, just as there is no national multiplication table; what is national is no longer science.

Note-Book of Anton Chekhov
(p. 18)

Genesis 22:17
I will multiply thy seed as the stars of the heaven, and as the sand which is upon the seashore.

The Bible

Goethe, Johann Wolfgang von
This you must ken!
From One make ten,
And two let be,
Make even three,
Then rich you'll be.
Skip o'er four!
From five and six,
The Witch's tricks,
Make seven and eight,
'Tis finished straight;
And nine is one,
And ten is none,
That is the witch's one-time-one!

Faust
The First Part
Witch's Kitchen, l. 2540–51

Jones, Cyrano
Twice nothing is still nothing.

Star Trek
The Trouble with Tribbles

Lichtenberg, Georg Christoph
If an angel were to tell us about his philosophy, I believe many of his statements might well sound like '2 × 2 = 13'.

Quoted in Franz H. Mautner and Henry Hatfield
The Lichtenberg Reader (p. 31)

Melrose, A.R.
Twy-stymes, *noun*: **1** arithmetic if it has to do with the number two {2} {which is always a good number to have when doing anything}. **2** multiplication table of the number two {2}.

The Pooh Dictionary

Zamyatin, Yevgeny
The multiplication table is wiser and more absolute than the ancient God: it never—do you realize the full meaning of the word?—it never errs.

We
Twelfth Entry (p. 59)

And there are no happier figures than those which live according to the harmonious, eternal laws of the multiplication table.

We
Twelfth Entry (p. 59)

NOTATION

Cajori, Florian
The miraculous powers of modern calculation are due to three inventions: the Arabic Notation, Decimal Fractions and Logarithms.

History of Mathematics (p. 161)

Dieudonné, Jean
This difficulty lead very gradually to the recognition of the need for a *shorthand* to make the sequence of operations easily comprehensible: here we have the problem of *notation*, which crops up again after every introduction of new objects, and which will probably never cease to torment mathematicians.

Mathematics—The Music of Reason (p. 49)

NUMBERS

Aeschylus
. . . but utterly without knowledge
Moiled, until I the rising of the stars
Showed them, and when they set, Though much obscure.
Moreever, number, the most excellent
Of all inventions, I for them devised,
And gave them writing that retaineth all,
The serviceable mother of the Muse.

The Plays of Aeschylus
Prometheus Bound
457

Asimov, Isaac
Human beings are very conservative in some ways and virtually never change numerical conventions once they grow used to them. They even come to mistake them for laws of nature.

Foundation and Earth (p. 376)

Auster, Paul
I've dealt with numbers all my life, of course, and after a while you begin to feel that each number has a personality of its own. A twelve is very different from a thirteen, for example. Twelve is upright, conscientious, intelligent, whereas thirteen is a loner, a shady character who won't think twice about breaking the law to get what he wants. Eleven is tough, an outdoorsman who likes tramping through woods and scaling mountains; ten is rather simpleminded, a bland figure who always does what he's told; nine is deep and mystical, a Buddha of contemplation . . . Numbers have souls, and you can't help but get involved with them in a personal way.

The Music of Chance
Chapter 4 (p. 73)

Bacon, Francis

. . . insomuch as we see in the schools both of Democritus and of Pythagoras, that the one did ascribe figure to the first seeds of things, and the other did suppose numbers to be the principles and originals of things.

The Advancement of Learning
Second Book, Chapter VIII, section 1

Berry, Daniel M.
Yavne, Moshe

In the beginning, everything was void, and J.H.W.H. Conway began to create numbers. Conway said, "Let there be two rules which bring forth all numbers large and small. This shall be the first rule: Every number corresponds to two sets of previously created numbers, such that no member of the left set is greater than or equal to any member of the right set. And the second rule shall be this: One number is less than or equal to another number if and only if no member of the first number's left set is greater than or equal to the second number, and no member of the second number's right set is less than or equal to the first number." And Conway examined these two rules he had made, and behold! they were very good.

Mathematics Magazine
The Conway Stones: What the Original Hebrew May Have Been
Volume 49, Number 4, September 1976 (p. 208)

Begley, Sharon

. . . number theory. It is a field of almost pristine irrelevance to everything except the wondrous demonstration that pure numbers, no more substantial than Plato's shadows, conceal magical laws and orders that the human mind can discover after all.

Newsweek
New Answer for an Old Question
5 July, 1993 (p. 53)

Bell, Eric T.

The theory of numbers is the last great uncivilized continent of mathematics. It is split up into innumerable countries, fertile enough in themselves, but all the more or less indifferent to one another's welfare and without a vestige of a central, intelligent government. If any young Alexander is weeping for a new world to conquer, it lies before him.

The Queen of the Sciences
Chapter VII (p. 91)

The algebraic numbers are spotted over the plane like stars against a black sky; the dense blackness is the firmament of the transcendentals.

Men of Mathematics (p. 569)

The next fundamental assumption of the Pythagoreans lies much deeper, so deep in fact that civilized man can scarcely hope to fetch it up to the full light of reason. Odd numbers are male; even numbers, female. We can only ask why, expecting no answer except possibly a hesitant allusion to a vestigial phallicism or a forgotten Orphism.

The Magic of Numbers (p. 155)

Bing, Ilse
numbers
supposedly
direct us
in a precise way
yet in certain combinations
numbers become
elusive
and their only answer
in the many answers
to our search
is
a question

Indeterminate Numbers
Quoted in Ernest Robson and Jet Wimp
Against Infinity

Borel, Émile
One grain of wheat does not constitute a pile, nor do two grains, nor three and so on. On the other hand, everyone will agree that a hundred million grains of wheat do form a pile. What, then, is the threshold number? Can we say that 325,647 grains of wheat do not form a pile, but that 325,648 grains do? If it is impossible to fix a threshold number, it will also be impossible to know what is meant by a pile of wheat; the words can have no meaning, although, in certain extreme cases, everybody will agree about them.

Probability and Certainty
Chapter 8 (p. 98)

Bridgman, Percy William
Nature does not count nor do integers occur in nature. Man made them all, integers and all the rest, Kroneker to the contrary notwithstanding.

The Way Things Are (p. 100)

Browning, Elizabeth Barrett
How do I love thee?
Let me count the ways.

The Complete Poetical Works of Elizabeth Barrett Browning
Sonnets from the Portuguese, XLIII

Burke, Edmund
The number is certainly the cause. The apparent disorder augments the grandeur.

> *On the Sublime and the Beautiful*
> Part II. Sec. XIII

Comte, Auguste
There is no inquiry which is not finally reducible to a question of Numbers; for there is none which may not be conceived of as consisting in the determination of quantities by each other, according to certain relations.

> *The Positive Philosophy*
> Volume I
> Book I, Chapter I (pp. 42–3)

Daniel 5:25
. . . *Mene, Mene, Tekel* and *Pharsin*
[Numbered, Numbered, Weighed, Divided]

> *The Bible*

Dunn, Alan
Does 63592187653.4215400532 look right to you?

> Caption on cartoon
> *Saturday Review*
> 1955

Ecclesiastes 1:2
The sand of the sea and the raindrops, and the days of eternity, who can assess them? The height of the sky and the breadth of the earth, and the depth of the abyss, who can probe them?

> *The Bible*

Eco, Umberto
With numbers you can do anything you like. Suppose I have the sacred number 9 and I want to get a number 1314, date of the execution of Jacques de Molay—a date dear to anyone who, like me, professes devotion to the Templar tradition of knighthood. What do I do? I multiply nine by one hundred and forty-six, the fateful day of destruction of Carthage. How do I arrive at this? I divided thirteen hundred and fourteen by two, three, et cetra, until I found a satisfying date. I could have also divided thirteen hundred and fourteen by 6.28, the double of 3.14, and I would have got two hundred and nine. That is the year Attalus I, king of Pergamon joined the anti-Macedonian League. You see?

> *Foucault's Pendulum*
> Chapter 48 (pp. 241–2)

Euripides
Numbers are a fearful thing, and joined to craft a desperate foe.

The Plays of Euripides
Hecuba
l. 884

Fabilli, Mary
What would I do
without numbers?
A 7 there and a 3 here,
days in a month
months in a year
AD and BC
and all such symbols.

Quoted in Ernest Robson and Jet Wimp
Against Infinity
Numbers

Ginsey, Gurney
There are numbers I don't trust . . . take
One—too proud, too pointed, much too
Sure that all begins with him—and
Two, that sits cross-legged on the
Path and sneers as though he knows all secrets . . .

Mathematics Magazine
Numbers (p. 168)
Volume 38, Number 3, May 1965

Hales, Stephen
. . . since we are assured that the all-wise Creator has observed the most
exact proportions, of *number, weight, and measure,* in the make of all things;
the most likely way therefore, to get any insight into the nature of these
parts of the creation, which come within our observation, must in all
reason be to number, weigh and measure.

Vegetable Staticks
Introduction (p. xxxi)

Hardy, Godfrey Harold
The elementary theory of numbers should be one of the very best subjects
for early mathematical instruction. It demands very little previous
knowledge; its subject matter is tangible and familiar; the processes
of reasoning which it employs are simple, general and few; and it is
unique among the mathematical sciences in its appeal to natural human
curiosity. A month's intelligent instruction in the theory of numbers

ought to be twice as instructive, twice as useful, and at least ten times as entertaining as the same amount of "calculus for engineers."

Bulletin of the American Mathematical Society
An Introduction to the Theory of Numbers
Volume 35, 1929 (p. 818)

Hofstadter, Douglas
People enjoy inventing slogans which violate basic arithmetic but which illustrate "deeper" truths, such as "1 and 1 make 1" (for lovers), or "I plus 1 plus 1 equals 1" (the Trinity) . . . Two raindrops running down a window merge; does one plus one make one? A cloud breaks up into two clouds—more evidence for the same? . . . Numbers as realities misbehave.

Gödel, Escher, Bach: An Eternal Golden Braid (p. 56)

Holmes, Oliver Wendell
Dr. Hooke, the famous English mathematician and philosopher, made a calculation of the number of separate ideas the mind is capable of entertaining, which he estimated as 3,155,760,000.—Haler, *Elementa Physiologiae*, vol. v. p. 547.

Pages from an Old Volume of Life (p. 274)

Huxley, Aldous
A million million spermatozoa,
  All of them alive:
Out of their cataclysm but one poor Noah
  Dare hope to survive.
And among that billion minus one
  Might have chanced to be
Shakespeare, another Newton, a new Donee—
  But the One was Me.

Stories, Essays, and Poems
Fifth Philosopher's Song

Jacobi, Karl G.J.
The God that reigns in Olympus is Number Eternal.

Quoted in Tobias Dantzig
Number: The Language of Science (p. 179)

Jevons, W. Stanley
Number is but another name for diversity.

The Principles of Science (p. 156)

Johnson, Samuel
Round numbers, she said, are always false.

Quoted in Mrs. Piozzi, Richard Cumberland, Bishop Percy and other's book
Johnsoniana
Apothegms, Sentiment, Opinions, &c

Juster, Norton

"How terribly confusing," he cried. "Everything here is called exactly what it is. The triangles are called triangles, the circles are called circles, and even the same numbers have the same name. Why, can you imagine what would happen if we named all the twos Henry or George or Robert or John or lots of other things? You'd have to say Robert plus John equals four, and if the four's name were Albert, things would be hopeless."

The Phantom Tollbooth (pp. 173–4)

Kaminsky, Kenneth

How poor were we? Why, we were so poor we only had imaginary numbers to play with.

Mathematics Magazine
Professor Fogelfroe
Volume 69, Number 4, October 1996 (p. 303)

Klein, William

Numbers are friends, for me, more or less. It doesn't mean the same for you, does it— 3,844? For you it's just a three and an eight and a four and a four. But I say, "Hi 62 squared."

Quoted in Oliver Sacks
The Man Who Mistook His Wife for a Hat and Other Clinical Tales
The Twins (pp. 198–9)

Kronecker, Leopold

Die ganzen Zahlen hat Gott gemacht, andere ist Menschenwerk.
[God created the natural numbers; everything else is man's handiwork.]

Jahresberichte der Deutschen Mathematiker Vereinigung
Book 2 (p. 19)

Number theorists are like lotus-eaters—having once tasted of this food they can never give it up.

Quoted in H. Eves
Mathematical Circles Squared
302°

Leibniz, Gottfried Wilhelm

Some things cannot be weighed, as having no force and power; some things cannot be measured, by reason of having no parts; but there is nothing which cannot be numbered.

Quoted in G. Frege
The Foundations of Arithmetic (p. 31)

There is an old saying that *God* created everything according to weight, measure and number.

Leibniz: Philosophical Papers and Letters
Volume I
On the General Characteristic (pp. 339–40)

Lewis, Gilbert Newton
. . . when we analyze the highly refined concept of space used by mathematicians we find it to be quite similar to the concept of number.

The Anatomy of Science (p. 29)

Lichtenberg, Georg Christoph
People don't like to choose lot #1 in a lottery. 'Choose it,' Reason cries loudly. 'It has as good a chance of winning the 12,000 thalers as any other.' 'In Heaven's name don't choose it,' a *je ne sais quoi* whispers. 'There's no example of such little numbers being listed before great winnings.' And actually no one takes it.

Lichtenberg: Aphorisms & Letters
Aphorisms (p. 46)

Locke, John
For number applies itself to men, angels, actions, thoughts; everything that either doth exist, or can be imagined.

Essays Concerning Human Understanding
Book II, Chapter xvi, Section 1

Longfellow, Henry Wadsworth
"Tell me not, in mournful numbers,
Life is but an empty dream!"

The Poems of Longfellow
A Psalm of Life
Stanza 1

Lover, Samuel
"Then here goes another," says he, "to make sure,
For there's luck in odd numbers," says Rory O'More.

Rory O'More

Maxwell, James Clerk
Thus numbers may be said to rule the whole world of quantity, and the four rules of arithmetic may be regarded as the complete equipment of the mathematician.

Quoted in Eric T. Bell
Men of Mathematics (p. xv)

Nordine, K.
. . . this makes me think of, of the somebody who invented numbers . . . Wonder if he had much trouble inventing those numbers; wonder if he had trouble with—zero. Pretty hard to imagine nothing.

Wordjazz [sound recording]
Looks Like It's Going To Rain (1957)

Parker, F.W.

Number was born in superstition and reared in mystery . . . Numbers were once made the foundation of religion and philosophy, and the tricks of figures have had a marvelous effect on a credulous people.

Talks on Pedagogics (p. 64)

Paulos, John Allen

The mathematician G.H. Hardy was visiting his protégé, the Indian mathematician Ramanujan, in the hospital. To make small talk, he remarked that 1729, the number of the taxi which had brought him, was a rather dull number, to which Ramanujan replied immediately, "No, Hardy! It is a very interesting number. It is the smallest number expressible as the sum of two cubes in two different ways."

Innumeracy (p. 6)

Philolaus

All things which can be known have number; for it is not possible that without number anything can be either conceived or known.

Quoted in Carl B. Boyer
A History of Mathematics (p. 60)

Plato

SOCRATES. And all arithmetic and calculations have to do with number?

GLAUCON. Yes.

SOCRATES. And they appear to lead the mind towards truth?

GLAUCON. Yes, in a very remarkable manner.

The Republic
Book VII, Section 525

Pliny the Elder

Why do we believe that in all matters the odd numbers are more powerful?

Natural History
Volume VIII, Book XXVIII, sec. 23

Plotinus

Objects of sense are not unlimited and therefore the Number applying to them cannot be so. Nor is an enumerator able to number to infinity; though we double, multiply over and over again, we still end with a finite number . . .

The Six Enneads
Sixth Ennead VI.2

Pope, Alexander
As yet a child, nor yet a fool to fame,
I lisp'd in numbers, for the numbers came.

The Complete Poetical Works of POPE
An Epistle to Dr. Arbuthnot
l. 127

Proclus
Wherever there is number, there is beauty.

Quoted in Morris Kline
Mathematical Thought from Ancient to Modern Times (p. 131)

Proverb, Pennsylvania German
It doesn't depend on size, or a cow would catch a rabbit.

Pythagoreans
Number rules the universe.

Quoted in E.T. Bell
Men of Mathematics (p. xv)

Quetelet, Adolphe
The official documents would make it appear that, of the 100,000 men, 28,620 are of less height than 5 feet 2 inches: calculation gives only 26,345. Is it not a fair presumption, that the 2,275 men who constitute the difference of these numbers have been fraudulently rejected? We can readily understand that it is an easy matter to reduce one's height a half-inch, or an inch, when so great an interest is at stake as that of being rejected.

Letters addressed to H.R.H. the Grand Duke of Saxe Coburg and Gotha on
the Theory of Probabilities as applied to
the Moral and Political Sciences

The RAND Corporation
A Million Random Digits with 100,000 Normal Deviates

Title of Book

Russell, Bertrand
We are the finite numbers.
We are the stuff of the world.
Whatever confusion cumbers
The earth is by us unfurled.
We revere our master Pythagoras
And deeply despise every hag or ass.
Not Endor's witch nor Balaam's mount

We recognize as wisdom's fount.
But round and round in endless baller
We move like comets seen by Halley.
And honored by the immortal Plato
We think no later mortal great-o.
We follow the laws
Without a pause,
For we are the finite numbers.

The Collected Stories of Bertrand Russell
Nightmares of Eminent Persons
The Mathematician's Nightmare (p. 43)

Sandburg, Carl
He was born to wonder about numbers.

Harvest Poems 1910–1960
Number Man

Shakespeare, William
. . . I am ill at these numbers.

Hamlet, Prince of Denmark
Act II, Scene 2, l. 120

This is the third time; I hope good luck lies in odd numbers . . . There is a divinity in odd numbers, either in nativity, chance or death.

The Merry Wives of Windsor
Act V, Scene 1, l. 2

'Smith, Adam'
pseud. Gooeman, George J.W.
. . . those who live by numbers can also perish by them and it is a terrifying thing to have an adding machine write an epitaph, either way.

The Money Game
Chapter 7 (p. 84)

Sukoff, Albert
Huge numbers are commonplace in our culture, but oddly enough the larger the number the less meaningful it seems to be . . . Anthropologists have reported on the primitive number systems of some aboriginal tribes. The Yancos in the Brazilian Amazon stop counting at three. Since their word for 'three' is *'poettarrarorincoaroac,'* this is understandable.

Saturday Review of the Society
Lotsa Hamburgers
March 1973 (p. 6)

Thomson, William [Lord Kelvin]
When you can measure what you are speaking about and express it in numbers you know something about it; but when you cannot express it in numbers, your knowledge is of a meager and unsatisfactory kind.

Lecture to the Institute of Civil Engineering
3 May 1883
Popular Lectures and Addresses
Volume I (p. 73)

Unknown
Defendit numerus.
[There is safety in numbers.]

Nec Babylonios temptaris numeros.
[Don't trust random numbers.]

ELEVEN: The first number not used in a countdown.

Quoted in Richard Iannelli
The Devil's New Dictionary

Virgil
Uneven numbers are the gods' delight.

The Eclogues
VIII, l. 77

Waismann, Friedrich
Will anyone seriously assert that the existence of negative numbers is guaranteed by the fact that there exist in the world hot assets and cold, and debts? Shall we refer to these things in the structure of arithmetic? Who does not see that thereby an entirely foreign element enters into arithmetic, which endangers the pureness and clarity of its concepts?

Introduction to Mathematical Thinking
Chapter 2 (p. 15)

West, Nathaniel
Prayers for the condemned man's soul will be offered on an adding machine. Numbers, he explained, constitute the only universal language.

Miss Lonelyhearts
Miss Lonelyhearts and the Deadpan

OBSERVATION

Anscombe, F.J.
No observations are absolutely trustworthy.

Technometrics
Rejection of Outliers
Volume 2, 1960 (p. 124)

Aristotle
. . . while those whom devotion to abstract discussions has rendered unobservant of the facts are too ready to dogmatize on the basis of a few observations.

On Generation and Corruption
Book I, Chapter II

Aurelius, Marcus [Antoninus]
Do not regulate your opinion by the caprice of a man that treats you contemptuously and would force you to adopt his own ideas; but examine things carefully, and decide according to truth.

The Meditations of the Emperor Antoninus Marcus Aurelius
Book IV, Section 10

Darwin, Charles
Oh, he is a good observer, but he has no power of reasoning!

The Life and Letters of Charles Darwin
Volume I
Mental Qualities (p. 82)

Einstein, Albert
A man should look for what is, and not for what he thinks should be . . .

Quoted in Peter Michelmore
Einstein, Profile of the Man (p. 20)

Greer, Scott
. . . the link between observation and formulation is one of the most difficult and crucial in the scientific enterprise. It is the process

of interpreting our theory or, as some say, of "operationalizing our concepts". Our creations in the world of possibility must be fitted in the world of probability; in Kant's epigram, "Concepts without precepts are empty". It is also the process of relating our observations to theory; to finish the epigram, "Precepts without concepts are blind".

The Logic of Social Inquiry (p. 160)

Jonson, Ben
I do love to note and to observe . . .

Volpone
Act II, Scene 1, l. 100

Langer, Susanne K.
The faith of scientists in the power of mathematics is so implicit that their work has gradually become less and less observation, and more and more calculation. The promiscuous collection and tabulation of data have given way to a process of assigning possible meanings, merely supposed real entities, to mathematical terms, working out the logical results, and then staging certain crucial experiments to check the hypothesis against the actual, empirical results. But the facts which are accepted by virtue of these tests are not actually *observed* at all.

Philosophy in a New Key (pp. 19–20)

Lehmer, D.N.
Perhaps in some far distant century they may say, "Strange that those ingenious investigators into the secrets of the number system had so little conception of the fundamental discoveries that would later develop from them."

Scripta Mathematica
Hunting Big Game in the Theory of Numbers
Volume 1, 1932 (pp. 229–35)

Longair, Malcolm
Although by now a large amount of observational material is available, the implications of the observations are far from clear.

Contemporary Physics
Quasi-Stellar Radio Sources
Volume 8, Number 4, July 1967 (p. 357)

Lyttleton, R.A.
Observations are meaningless without a theory to interpret them.

Quoted in Charles-Albert Reichen
A History of Astronomy (p. 88)

Poincaré, Henri
. . . to observe is not enough. We must use our observations, and to do that we must generalize.

The Foundations of Science
Science and Hypothesis (p. 127)

Pope, Alexander
To observations which ourselves we make,
We grow more partial for th' observer's sake.

The Complete Poetical Works
Moral Essays
Epis. I, l. 11

Swift, Jonathan
That was excellently observ'd, say I, when I read a Passage in an Author, where his Opinion agrees with mine. When we differ, there I pronounce him to be mistaken.

Satires and Personal Writings
Thoughts on Various Subjects

Sylvester, James Joseph
Most, if not all, of the great ideas of modern mathematics have had their origin in observation.

Nature
A Plea for the Mathematician
Volume 1, 1869 (p. 238)

OPINIONS

Lippmann, Walter
True opinions can prevail only if the facts to which they refer are known; if they are not known, false ideas are just as effective as true ones, if not a little more effective.

Liberty and the News (pp. 64–5)

Locke, John
New opinions are always suspected, and usually opposed, without any other reason but because they are not already common.

An Essay Concerning Human Understanding
Dedicatory Epistle

Terence
. . . there are as many opinions as there are people . . .

Phormio
Act II, scene iv

Unknown
. . . opinions ought to count by weight rather than by number . . .

Quoted in James Joseph Sylvester
Collected Mathematical Works
Volume III
Address on Commemoration Day at Johns Hopkins University (p. 73)

ORDER

Huntington, E.V.
The fundamental importance of the subject of order may be inferred from the fact that all the concepts required in geometry can be expressed in terms of the concept of order alone.

The Continuum
Introduction (p. 2, fn *)

Picard, Émile
We no longer pretend to be able to grasp reality in a physical theory; we see in it rather an analytic or geometric mold useful and fertile for a tentative representation of phenomena, no longer believing that the agreement of a theory with experience demonstrates that the theory expresses the reality of things. Such statements have sometimes seemed discouraging; we ought rather to marvel that, with representations of things more or less distant and discolored, the human spirit has been able to find its way through the chaos of so many phenomena and to derive from scientific knowledge the ideas of beauty and harmony. It is no paradox to say that science puts order, at least tentative order, into nature.

Quoted in Lucienne Felix
The Modern Aspects of Mathematics (p. 31)

Poincaré, Henri
To obtain a result of real value, it is not enough to grind out calculations or to have a machine to put things in order; it is not order alone, it is unexpected order, which is worth while. The machine may gnaw on the crude fact, the soul of the fact will always escape it.

Science and Method
The Future of Mathematics (p. 32)

Russell, Bertrand

Dimensions, in geometry, are a development of order. The conception of a *limit*, which underlies all higher mathematics, is a serial conception. There are parts of mathematics which do not depend upon the notion of order, but they are very few in comparison with the parts in which this notion is involved.

Introduction to Mathematical Philosophy
The Definition of Order (p. 29)

The notion of continuity depends upon that of *order*, since continuity is merely a particular type of order.

Mysticism and Logic
Mathematics and Metaphysics (p. 91)

WE EXPECT A
BREAKTHROUGH
ANY TIME NOW!

When you walk in on a research mathematician and find him reclining with his feet up, gazing wistfully out the window, what you say is: "Sorry, I didn't know you were working." Because he probably is.

Jerry P. King – (See p. 160)

ORDINALS

Gass, Fredrick

Adam, did you find a good system for naming ordinals?

A: Ordinals? I thought you said "animals."

Mathematics Magazine
Constructive Ordinal Notation Systems
Volume 57, Number 3, May 1984 (p. 131)

PARABOLA

Allen, Woody
She wore a short skirt and a tight sweater and her figure described a set of parabolas that could cause cardiac arrest in a yak.

Getting Even (p. 139)

Frere, C.
Canning, B.
And first, the fair PARABOLA behold,
Her timid arms, with virgin blush, unfold!
Though, on one *focus* fixed, her eyes betray
A heart that glows with love's resistless sway;

Quoted in Charles Edmonds
Poetry of the Anti-Jacobin
The Loves of the Triangle
Canto II, l. 107–8

Shaw, George Bernard
Kneller: . . . take a sugar loaf and cut it slantwise, and you will get hyperbolas and parabolas, ellipses and ovals . . .

The Complete Plays of Bernard Shaw
In Good Kind Charles's Golden Days
Act I (p. 1358)

PARADOX

Bohr, Niels
How wonderful that we have met with a paradox. Now we have some hope of making progress.

Quoted in L.I. Ponomarev
The Quantum Dice (p. 75)

Bourbaki, Nicholas
There is no sharply drawn line between those contradictions which occur in the daily work of every mathematician, beginner or master of his craft, as a result of more or less easily detected mistakes, and the major paradoxes which provide food for logical thought for decades and sometimes centuries.

Quoted in Bryan H. Bunch
Mathematical Fallacies and Paradoxes (p. 38)

Eliot, George
Play not with paradoxes. That caustic which you handle in order to scorch others may happen to sear your own fingers and make them dead to the quality of things.

Felix Holt, The Radical
Chapter 13 (p. 131)

Gleason, Andrew
If we do not run into a paradox, we can probably save the structure of mathematics by patching it.

Quoted in Byron H. Bunch
Mathematical Fallacies and Paradoxes (p. 110)

Kasner, Edward
Newman, James R.
Perhaps the greatest paradox of all is that there are paradoxes in mathematics . . . because mathematics builds on the old but does not

discard it, because its theorems are deduced from postulates by the methods of logic, in spite of its having undergone revolutionary changes we do not suspect it of being a discipline capable of engendering paradoxes.

Mathematics and the Imagination (p. 193)

Rogers, Hartley, Jr.

It is a paradox in mathematics and physics that we have no good model for the teaching of models.

Quoted in Lynn Arthur Steen
Mathematics Tomorrow (p. 232)

Russell, Bertrand

Although this may seem a paradox; all exact science is dominated by the idea of approximation.

Quoted in Jefferson Hane Weaver
The World of Physics
Volume II (p. 22)

Smith, E.E.

With sufficient knowledge, any possible so-called paradox can be resolved . . .

Masters of the Vortex (p. 109)

Thomson, William [Lord Kelvin]

In science there are no paradoxes.

Quoted in S.P. Thompson
The Life of William Thomson Baron Kelvin of Largs (p. 833)

Wilde, Oscar

The way of paradoxes is the way of truth. To test Reality we must see it on the tight-rope. When the Verities become acrobats we can judge them.

The Picture of Dorian Gray
Chapter 3

PARALLELOGRAM

Moultrie, John
. . . forgetful of the claims
of curves and squares, and parallelograms,
Cones, angles, sines and cosines, ordinates,
Abscissae and the like.

The Dream of Life
III, Youth

Tolstoy, Leo
A countless number of free forces (for nowhere is man freer than during a battle, where it is a question of life and death) influence the course taken by the fight, and that course never can be known in advance and never coincides with the direction of any one force. If many simultaneously and variously directed forces act on a given body, the direction of its motion cannot coincide with any one of those forces, but will always be a mean—what in mechanics is represented by the diagonal of a parallelogram of forces. If in the descriptions given by historians, especially French ones, we find their wars and battles carried out in accordance with previously formed plans, the only conclusion to be drawn is that those descriptions are false.

War and Peace
Fourth book, Second part, Chapter 8

PERFECT NUMBER

Dickson, L.E.
Perfect numbers have engaged the attention of arithmeticians of every century of the Christian era.

History of the Theory of Numbers
Volume I (p. III)

PI

Beckmann, Petr
The digits beyond the first few decimal places are of no practical or scientific value. Four decimal places are sufficient for the design of the finest engines; ten decimals are sufficient to obtain the circumference of the earth to within a fraction of an inch if the earth were a smooth sphere . . .

A History of Pi (p. 100)

Carter, Harvey L.
'Tis a favorite project of mine
A new value of pi to assign.
  I would fix it at 3
    For it's simpler, you see,
Than 3 point 1 4 1 5 9.

Quoted in W.S. Baring-Gould
The Lure of the Limerick (p. 5)

Duffin, R.J.
God created the world and the integers, all in seven days. He then ordered two of his biotechnicians, James and Francis, to construct a genetic code for the fractional numbers. Moreover, they were to give special prominence to His favorite number, π.

The Mathematical Intelligencer
The Patron Saint of Mathematics
Volume 15, Number 1, 1993 (p. 52)

Graham, L.A.
Fiddle de dum, fiddle de dee,
A ring round the moon is π times D;
But if a hole you want repaired,
You use the formula πr^2.

Ingenious Mathematical Problems and Methods
Mathematical Nursery Rhyme No. 1

Little Jack Horner sat in a corner,
Trying to evaluate π.
  He disdained rule of thumb,
  Found an infinite sum,
And exclaimed "It's REAL, nary an I."

Ingenious Mathematical Problems and Methods
Mathematical Nursery Rhyme No. 9

Kac, Mark
Steinhaus, with his predilection for metaphors, used to quote a Polish proverb, *'Forturny kolem sie tocza' (Luck runs in circles)*, to explain why π, so intimately connected with circles, keeps cropping up in probability theory and statistics, the two disciplines which deal with randomness and luck.

Enigmas of Chance (p. 55)

I Kings 7:23
And he made a molten sea, ten cubits from the one brim to the other: it was round all about, and its height was five cubits: and a line of thirty cubits did compass it about.

The Bible

Morgan, Robert
The secret relationship
of line and circle, progress
and return, is always known,
transcendental and yet
a commonplace. And though
the connection is written
it cannot be written out
in full, never perfect, but
is exact and constant, is
eternal and everyday
as orbits of electrons,
chemical rings, noted here
in one brief sign as gateway
to completed turns and
the distance inside circles,
both compact and infinite.

Poetr
ɼ
Volume CLXI, Number 4, January 1993 (p. 204

Preston, Richard

Pi is not the solution to any equation built from a less than infinite series of whole numbers. If equations are trains threading the landscape of numbers, then no train stops at pi.

Quoted in Clifford A. Pickover
Keys to Infinity (p. 60)

The study of mathematics is like climbing up a steep and craggy mountain; when once you reach the top, it fully recompenses your trouble, by opening a fine, clear, and extensive prospect.

Tyron Edwards – (See p. 196)

PI MNEMONICS

Burr, E. Scott
Now I, even I, would celebrate
In rhymes inept the great
Immortal Syracusan, rivaled nevermore
Who in his wondrous lore
Passed on before.

The Physics Teacher
A Rhyming Π
February 1979 (p. 143)

Fadiman, Clifton
In order to remember the value of pi to thirty places of decimals you need only learn . . . the following passage by heart . . . The integer is purposely omitted.

I nunc, O Baili, Parnassum et desere rupen;
  Dic sacra Pieridum deteriora quadris!
Subsidium hoc ad vos, quamquam leve, fertur ab hymnis
  Quos dat vox Sophocli (non in utroque probrumst?)

[Will you dare any longer, to turn your back on Parnassus hill, telling us that the sacred rites of the Muses are less important than constructing squares?]

The Mathematical Magpie (p. 287)

Kamath, B.A.
Que j'aime à faire apprendre un
nombre utile aux sages
Immortel Archimède, artiste, ingénieur
Qui de ton jugement peut priser
la valeur.

The Physics Teacher
The end of Π
May 1979 (p. 343)

Kirkpatrick, Larry D.
Sir, I send a rhyme excelling
In sacred truth and rigid spelling
Numerical sprites elucidate
For me the lexicon's dull weighty

The Physics Teacher
Numerical Sprites Indeed
November 1978 (p. 591)

Unknown
Now I, even I, would celebrate
In rhymes inapt, the great
Immortal Syracusan, rivaled nevermore,
Who in his wondrous lore,
Passond on before,
Left men his guidance how to circles mensurate.

Literary Digest
More Mathematical Verse
Volume 32 (1906) (p. 84)
Contributed by A.C. Orr

'Tis a king, a giant intellect of mighty force.
The great immortal Syracuson arrests attention,
for by his geometry yond temple is raised,
land and sky measured. Let us quietly reverence.

Literary Digest
More Mathematical Verse
Volume 32 (1906) (p. 84)

Dir, o Held, o alter Philosoph, du Riesengenie!
Wie viele Tausende bewundern Geister
Himmlisch wie du und göttlich!
Noch reiner in Aeonen
Wird das uns strahlen
Wie im lichten Morgenrot!

In Petr Beckman
A History of Pi (p. 108)

Wie? O! Dies π
Macht ernstlich so vielen, viele Müh'!
Lernt immerhin, Jünglinge, leichte Verselein,
Wie so zum Beispiel dies dürfte zu merken sein!

In Petr Beckman
A History of Pi (p. 109)

POINT

Warner, Sylvia Townsend
He took out his pocket knife and whittled the end of the stick. Then he tried again.

"What is this?"

"A smaller hole."

"Point," said Mr. Fortune suggestively.

"Yes, I mean a smaller point."

"No, not quite. It is a point, but it is not smaller. Holes may be of different sizes, but no point is larger or smaller than another point."

Mr. Fortune's Maggot (p. 164)

. . . if a given point were not in a given place it would not be there at all.

Mr. Fortune's Maggot (p. 166)

POSTULATE

Frere, C.
Canning, B.
To *you* no POSTULATES prefer their claim,
No Ardent AXIOMS *your* dull souls inflame;
For *you* no TANGENTS touch, no ANGLES meet,
No CIRCLES join in osculation sweet!

Quoted in Charles Edmonds
Poetry of the Anti-Jacobin
The Loves of the Triangle
Canto I, l. 7–10

Russell, Bertrand
The method of 'postulating' what we want has many advantages; they are the same as the advantages of theft over honest toil. Let us leave them to others and proceed with our honest toil.

Introduction to Mathematical Philosophy (p. 71)

PRAYER

Adams, Douglas
Protect me from knowing what I don't need to know. Protect me from even knowing that there are things to know that I don't know. Protect me from knowing that I decided not to know about the things that I decided not to know about. Amen.

Mostly Harmless
Chapter 9 (p. 84)

Lord, lord, lord. Protect me from the consequences of the above prayer. Amen.

Mostly Harmless
Chapter 9 (p. 84)

Huxley, Thomas
God give me the strength to face a fact though it slay me.

Quoted by George Seldes in
The Great Quotations (p. 344)

Lederman, Leon
Dear Lord, forgive me the sin of arrogance, and Lord, by arrogance I mean the following . . .

Quoted by John D. Barrow
The Artful Universe (p. 31)

Lewis, Sinclair
God give me the unclouded eyes and freedom from haste. God give me quiet and relentless anger against all pretense and all pretentious work and all work left slack and unfinished. God give me a restlessness whereby I may neither sleep nor accept praise till my observed results equal my calculated results or in pious glee I discover and assault my error. God give me the strength not to trust to God.

Arrowsmith
Chapter XXVI, Section II (p. 292)

Tukey, John W.
The physical sciences are used to "praying over" their data, examining the same data from a variety of points of view. This process has been very rewarding, and has led to many extremely valuable insights. Without this sort of flexibility, progress in physical science would have been much slower. Flexibility in analysis is often to be had honestly at the price of a willingness not to demand that what has *already* been observed shall establish, or prove, what analysis *suggests*. In physical science generally, the results of praying over the data are thought of as something to be put to further test in another experiment, as indications rather than conclusions.

The Annals of Mathematical Statistics
The Future of Data Analysis
Volume 33, Number 1, March 1962 (p. 46)

Unknown
Grant, oh God, Thy benedictions
On my theories, predictions
Lest the facts, when verified,
Show Thy servant to have lied.

Proceedings of the Chemical Society
January 1963 (pp. 8–10)

God grant that no one else has done
  The work I want to do,
Then give me the wit to write it up
  In decent English, too.

Applied Optics
Of Optics and Opticists
Volume 8, Number 2, February 1969 (p. 273)

PRECISION

Thompson, Sir D'Arcy Wentworth
Numerical precision is the very soul of science.

On Growth and Form
Chapter 1

The student must be careful in calculations involving the decimal point to put it in its exact place, neither too much to the right nor too much to the left.
Hilaire Belloc – (See p. 23)

PRIME

Auster, Paul
Prime numbers. It was all so neat and elegant. Numbers that refuse to cooperate, that don't change or divide, numbers that remain themselves for all eternity.

The Music of Chance
Chapter 4 (pp. 73–4)

Euler, Leonhard
Mathematicians have tried in vain to this day to discover some order in the sequence of prime numbers, and we have reason to believe that it is a mystery into which the human mind will never penetrate. To convince ourselves, we have only to cast a glance at tables of primes, which some have taken the trouble to compute beyond a hundred thousand, and we should perceive at once that there reigns neither order nor rule.

Collected Works
Serial 1, Volume 2 (p. 241)

Gardner, Martin
The primes . . . those exasperating, unruly integers that refuse to be divided evenly by any integers except themselves and one.

Quoted in Eli Maor
To Infinity and Beyond: A Cultural History of the Infinite (p. 21)

Queneau, Raymond
When One made love to Zero
spheres embraced their arches
and prime numbers caught their breath . . .

Pounding the Pavement, Beating the Bush, and Other Pataphysical Poems
Sines

PRINCIPLE

Pólya, George
This principle is so perfectly general that no particular application is possible.

How to Solve It (p. 181)

PROBABILITY

Arago
The calculus of probabilities, when confined within just limits, ought to interest, in an equal degree, the mathematician, the experimentalist, and the statesman. From the time when Pascal and Fermat established its first principles, it has rendered, and continues daily to render, services of the most eminent kind. It is the calculus of probabilities, which, after having suggested the best arrangements of the tables of populations and mortality, teaches us to deduce from those numbers, in useful character; it is the calculus of probabilities which alone can regulate justly the premiums to be paid for assurances; the reserve funds for the disbursements of pensions, annuities, discounts, etc. It is under its influence that lotteries and other shameful snares cunningly laid for avarice and ignorance have definitely disappeared.

Smithsonian Report
Eulogy on Laplace
1874 (p. 164)

Arbuthnot, John
The Reader may here observe the Force of Numbers, which can be successfully applied, even to those things, which one would imagine are subject to no Rules. There are very few things which we know, which are not capable of being reduc'd to a Mathematical Reasoning, and when they cannot, its a Sign our Knowledge of them is very small and confus'd; And where mathematical reasoning can be had, its as great folly to make use of any other, as to grope for a thing in the dark, when you have a Candle standing by you. I believe the Calculation of the Quantity of Probability might be improved to a very useful and pleasant Speculation, and applied to a great many Events which are accidental, besides those of Games; . . .

Of the Laws of Chance
Preface

Boole, George

Probability is expectation founded upon partial knowledge. A perfect acquaintance with *all* the circumstances affecting the occurrence of an event would change expectation into certainty, and leave neither room nor demand for a theory of probabilities.

Collected Logical Works
Volume II
An Investigation of the Law of Thought
Chapter XVI (p. 258)

Borel, Émile

Probabilities must be regarded as analogous to the measurement of physical magnitudes; that is to say, they can never be known exactly, but only within certain approximation.

Probabilities and Life
Introduction (pp. 32–3)

Cohen, John

Unlike almost all mathematics, I agree completely with your statement that every probability evaluation is a probability evaluation, that is, something to which it is meaningless to apply such attributes as *right*, *wrong*, *rational*, etc.

Chance, Skill, and Luck
Chapter 2, Part 1 (p. 28)

Crofton, M.W.

The mathematical theory of probability is a science which aims at reducing to calculation, where possible, the amount of credence due to propositions or statements, or to the occurrence of events, future or past, more especially as contingent or dependent upon other propositions or events the probability of which is known.

Encyclopaedia Britannica
9th Edition
Probability

Feller, William

Probability is a mathematical discipline with aims akin to those, for example, of geometry or analytical mechanics. In each field we must carefully distinguish three aspects of the theory: (a) the formal logical content, (b) the intuitive background, (c) the applications. The character, and the charm, of the whole structure cannot be appreciated without considering all three aspects in their proper relation.

An Introduction to Probability Theory and Its Applications (p. 1)

All possible "definitions" of probability fall short of the actual practice.

An Introduction to Probability Theory and Its Applications (p. 19)

Kolmogorov, Andrey N.
The theory of probability as mathematical discipline can and should be developed from axioms in exactly the same way as Geometry and Algebra.

Foundations of the Theory of Probability
Chapter 1
Elementary Theory of Probability (p. 1)

Kyburg, H.E., Jr.
Smokler, H.E.
. . . there is no problem about probability: it is simply a non-negative, additive set function, whose maximum value is unity.

Studies in Subjective Probability (p. 3)

Peirce, Charles Sanders
This branch of mathematics [probability] is the only one, I believe, in which good writers frequently get results entirely erroneous.

Writings of Charles Sanders Peirce
Volume 3 (p. 279)

Woodward, Robert S.
The theory of probabilities and the theory of errors now constitute a formidable body of knowledge of great mathematical interest and of great practical importance. Though developed largely through the applications to the more precise sciences of astronomy, geodesy, and physics, their range of applicability extends to all the sciences; and they are plainly destined to play an increasingly important role in the development and in the applications of the sciences of the future. Hence their study is not only a commendable element in a liberal education, but some knowledge of them is essential to a correct understanding of daily events.

Probability and Theory of Errors
Preface

PROOF

Bell, Eric T.
There is a sharp disagreement among competent men as to what can be proved and what cannot be proved, as well as an irreconcilable divergence of opinions as to what is sense and what is nonsense.

Debunking Science

Beveridge, W.I.B.
Generalisations can never be *proved*.

The Art of Scientific Investigation (p. 88)

Blake, William
What is now proved was once only imagined.

BLAKE: The Complete Poems
The Marriage of Heaven and Hell

Bourbaki, N.
Indeed every mathematician knows that a proof has not been "understood" if one has done nothing more than verify step by step the correctness of the deductions of which it is composed and has not tried to gain a clear insight into the ideas which have led to the construction of this particular chain of deductions in preference to every other one.

Quoted in Douglas M. Campbell and John C. Higgins
Mathematics: People, Problems, Results
Volume III (p. 25)

Buchanan, Scott
The best proofs in mathematics are short and crisp like epigrams, and the longest have swings and rhythms that are like music.

Poetry and Mathematics (p. 36)

de Morgan, Augustus
Would Mathematicals—forsooth—
If true, have failed to prove truth?
Would not they—if they could—submit
Some overwhelming proofs of it?

A Budget of Paradoxes
Volume II (p. 21)

Eddington, Sir Arthur Stanley
Proof is an idol before which the mathematician tortures himself.

Quoted in Clifford A. Pickover
Keys to Infinity (p. 2)

Euclid
Quod erat demonstrandum (Q.E.D.)
[Which was to be proved.]

The Thirteen Books of Euclid's Elements
Element I, Proposition 5

Evans, Bergen
"You can't prove it *isn't* so!" is as good as Q.E.D. in folk logic.

The Natural History of Nonsense
A Tale of a Tub

Hilbert, David
... it is an error to believe that rigor in the proof is the enemy of simplicity
...

Bulletin of the American Mathematical Society
Hilbert: Mathematical Problems
Volume 8, 2nd series, October 1901–July 1902 (p. 441)

Hoyle, Fred
What constitutes proof in one generation is not the same thing as proof in another.

Of Men and Galaxies (p. 17)

Kline, Morris
Much research for new proofs of theorems already correctly established is undertaken simply because the existing proofs have no aesthetic appeal. There are mathematical demonstrations that are merely convincing; to use a phrase of the famous mathematical physicist, Lord Rayleigh, they 'command assent.' There are other proofs 'which woo and charm the intellect. They evoke delight and an overpowering desire to say, Amen, Amen.' An elegantly executed proof is a poem in all but the form in which it is written.

Mathematics in Western Culture (p. 470)

Lohwater, A.J.
The Devil said to Daniel Webster, "Set me a task I can't carry out, and I'll give you anything in the world you may ask for."

Daniel Webster: "Fair enough. Prove that for n greater than 2, the equation $a^n + b^n = c^n$ has no non-trivial solution in integers."

They agreed on a three-day period for the labor, and the Devil disappeared.

At the end of the three days the Devil presented himself, haggard, jumpy, biting his lip. Daniel Webster said to him, "Well—how did you do at my task? Did you prove the theorem?"

"Eh? No . . . no, I haven't proved it."

"Then I can have whatever I ask for? Money? The Presidency?"

"What? Oh, *that*—of course. But listen! If we could just prove the following two lemmas—"

> Quoted in Clifton Fadiman
> *The Mathematical Magpie* (p. 261)

Manin, Yu I.
. . . a good proof is one that makes us wiser.

> *A Course in Mathematical Logic* (p. 51)

Pearson, Karl
. . . we must remember that because a proposition has not yet been proved, we have no right to infer that its converse must be true.

> *The Grammar of Science*
> Cause and Effect—Probability (p. 150)

Pólya, George
If you have to prove a theorem, do not rush. First of all, understand fully what the theorem says, try to see clearly what it means. Then check the theorem; it could be false. Examine the consequences, verify as many particular instances as are needed to convince yourself of its truth. When you have satisfied yourself that the theorem is true, you start proving it.

> *How to Solve It* (p. 181)

Proverb, English
The proof of the pudding is in the eating.

Shakespeare, William
Be sure of it: give me the ocular proof . . .

> *Othello: Moor of Venice*
> Act III, Scene III, l. 360

Stewart, Ian
Proofs knit the fabric of mathematics together, and if a single thread is weak, the entire fabric may unravel.

Nature's Numbers (p. 45)

Sylvester, James Joseph
It always seems to me absurd to speak of a complete proof, or of a theorem being rigorously demonstrated. An incomplete proof is no proof, and a mathematical truth not rigorously demonstrated is not demonstrated at all.

The Collected Mathematical Papers of James Joseph Sylvester
Volume 2 (p. 200)

Tymoczko, Thomas
A proof is a construction that can be looked over, reviewed, verified by a rational agent. We often say that a proof must be perspicuous or capable of being checked by hand. It is an exhibition, a derivation of the conclusion, and it needs nothing outside itself to be convincing. The mathematician *surveys* the proof in its entirety and thereby comes to *know* the conclusion.

Journal of Philosophy
The Four Color Problems
Volume 76, 1979

Unknown
Sex and drugs? They're nothing compared with a good proof!

Mathematical and Scientific Quotes from Cambridge
The Internet

Proof of Thm. 6.2 is trivial from Thm. 6.9.

Mathematical and Scientific Quotes from Cambridge
The Internet

You don't want to prove theorems that are false.

Mathematical and Scientific Quotes from Cambridge
The Internet

An obscure proof which I managed to present in an obscure way.

Source unknown

White, Arthur
A teacher once, having some fun,
In presenting that two equals one,
  Remained quite aloof
  From his rigorous proof;
But his class was convinced and undone.

Mathematical Magazine
Volume 64, Number 2, April 1991 (p. 91)

Zemanian, Armen H.
The usual techniques for proving things are often inadequate because they are merely concerned with truth. For more practical objectives, there are other powerful—but generally unacknowledged—methods.

Proof of Blatant Assertion:

Use words and phrases like "clearly . . . ," "obviously . . . ," "it is easily shown that . . . ," and "as any fool can plainly see . . . "

Proof by Seduction:

If you will just agree to believe this, you might get a better final grade.

Proof by Intimidation:

You better believe this if you want to pass the course.

Proof by Interruption:

Keep interrupting until your opponent gives up.

Proof by Misconception:

An example of this is the Freshman's Conception of the Limit Process: "2 equals 3 for large values of 2." Once introduced, any conclusion is reachable.

Proof by Obfuscation:

A long list of lemmas is helpful in this case—the more, the better.

Proof by Confusion:

This is a more refined form of proof by obfuscation. The long list of lemmas should be arranged into circular patterns of reasoning—and perhaps more baroque structures such as figure-eights and fleurs-de-lis.

Proof by Exhaustion:

This is a modification of an inductive proof. Instead of going to the general case after proving the first one, prove the second case, then the third, then the fourth, and so on—until a sufficiently large n is achieved whereby the nth case is being propounded to a soundly sleeping audience.

The Physics Teacher
Appropriate Proof Techniques
Volume 32, Number 5, May 1994 (p. 287)

PYTHAGOREANS

Aristotle
... the so-called Pythagoreans, who were the first to take up mathematics, not only advanced this subject, but having been brought up in it, they thought its principles of mathematics were the principles of all things.

Metaphysics
Book 1, Chapter 5

Browne, Sir Thomas
I have often admired the mystical way of Pythagoras, and the secret magic of numbers.

Religio Medici
Part 1, Section 11

Russell, Bertrand
With equal passion I have sought knowledge. I have wished to understand the hearts of men. I have wished to know why the stars shine. And I have tried to apprehend the Pythagorean power by which number holds sway above the flux. A little of this, but not much, I have achieved.

The Autobiography of Bertrand Russell
Prologue (pp. 3–4)

Scarecrow
The sum of the square roots of any two sides of an isosceles triangle is equal to the square root of the remaining side.

The Wizard of Oz
Misstating the Pythagorean Theorem when he received the diploma
from the wizard
1936

313

van der Waerden, B.L.

The problem of approximating to the ratio of the diagonal and the side [of a square] by means of rational numbers was proposed and solved by the Babylonians. But the Pythagoreans carried this old problem infinitely farther than the Babylonians. They found a whole set of approximations of indefinitely increasing accuracy; moreover they developed a scientific theory concerning these approximations and they proved the general proposition by complete induction. Again and again it becomes apparent that there were excellent number theoreticians in the Pythagorean school.

Science Awakening (p. 127)

I WARNED HIM –
GIVE UP MATHEMATICS
AND YOUR HOUSE
WILL FALL DOWN....!

Therefore, O students, study mathematics, and do not build without foundations.

Leonardo da Vinci – (See p. 192)

REASONING

Barnett, P.A.
. . . the reasoning of mathematics is a type of perfect reasoning.

Common Sense in Education and Teaching (p. 222)

Beveridge, W.I.B.
How easy it is for unverified assumptions to creep into our reasoning unnoticed!

The Art of Scientific Investigation (p. 87)

Eldridge, Paul
Reason is the shepherd trying to corral life's vast flock of wild irrationalities.

Maxims for a Modern Man

Hubbard, Elbert
REASON: The arithmetic of the emotions.

The Roycroft Dictionary (p. 126)

Newton, Sir Isaac
My Design in this Book is not to explain the Properties of Light by Hypotheses, but to propose and prove by Reason and Experiments: In order to which I shall premise the following Definitions and Axioms.

Opticks
Book One, Part I

Romanoff, Alexis L.
Reasoning goes beyond the analysis of facts.

Encyclopedia of Thoughts
Aphorisms
1973

Shakespeare, William

His reasons are as two grains of wheat hid in two bushels of chaff: you shall seek all day ere you find them, and when you have them, they are not worth the search.

The Merchant of Venice
Act I, Scene 1, l. 115

Wells, H.G.

"It's against reason," said Filby.
"What reason? said the Time Traveler.

28 Science Fiction Stories of H.G. Wells
The Time Machine

Whitehead, Alfred North

The art of reasoning consists in getting hold of the subject at the right end, of seizing on the few general ideas that illuminate the whole, and of persistently organizing all subsidiary facts round them. Nobody can be a good reasoner unless by constant practice he has realized the importance of getting hold of the big ideas and hanging on to them like grim death.

Presidential Address to the London Branch of the
Mathematical Association, 1914
Quoted in W.W. Sawyer
Prelude to Mathematics (p. 183)

RECTANGLE

Frere, C.
Canning, B.
Alas! that partial Science should approve
The sly RECTANGLE'S too licentious love!

Quoted in Charles Edmonds
Poetry of the Anti-Jacobin
The Loves of the Triangle
Canto II, l. 75–6

RECURSION

Kelly-Bootle, Stan
recursive *adj. See* RECURSIVE.

The Devil's DP Dictionary

Papert, Seymour
Of all ideas I have introduced to children, recursion stands out as the one idea that is particularly able to evoke an excited response.

Mindstorm (p. 71)

Unknown
To iterate is human, to recurse divine.

Source unknown

REFEREES

Lipschitz, R.

Sir,

It is sometimes a matter of wonder, to us in Hades, that what we had believed to be our best work remains buried under thick layers of dust in your libraries, while the very talented young men in the mathematical world of the present day strive manfully against problems which are by no means as novel as they think . . . Unfortunately, it appears that there is now in your world a race of vampires, called referees, who clamp down mercilessly upon mathematicians unless they know the right passwords. I shall do my best to modernize my language and notions, but I am well aware of my shortcomings in that respect; I can assure you, at any rate, that my intentions are honourable and my results invariant, probably canonical, perhaps even functional. But please allow me to assume that the characteristic is not 2.

Annals of Mathematics
2nd Series
Letter to the Editor
Volume 69, 1959 (pp. 247–8)

RELATIONS

Keyser, Cassius J.
To be is to be related.

Mole Philosophy and Other Essays (p. 94)

A mathematical theory is not to be considered complete until you have made it so clear that you can explain it to the first man whom you meet on the street.

Unknown – (See p. 143)

RESEARCH

Bates, Marston
Research is the process of going up alleys to see if they are blind.

Quoted in Jefferson Hane Weaver
The World of Physics
Volume II (p. 63)

Dodgson, Charles L.
Now this field of Mathematical research, with all its wealth of hidden treasures, is all too apt to yield nothing to our research: for it is haunted by certain *ignes fatui*—delusive phantoms, that float before us, and seem so fair, and are *all but* in our grasps, so nearly that it never seems to need more than *one* step further, and the prize shall be ours! Alas for him who has been turned aside from real research by one of these spectres—who has found a music in its mocking laughter—and who wastes his life and energy in the desperate chase!

A New Theory of Parallels (p. xvi)

Green, Celia
The way to do research is to attack the facts at the point of greatest astonishment.

The Decline and Fall of Science
Aphorisms (p. 1)

Kline, Morris
Mathematical research is also becoming highly professionalized in the worst sense of that term. Research performed voluntarily and sincerely by devoted souls, research as a relish of knowledge, is to be welcomed even if the results are minor. But hothouse-grown research, which crowds the journals and promotes only promotion, is a drag on science.

Why The Professor Can't Teach
The Nature of Current Mathematical Research

Lasker, Albert D.
"Research," he said, "is something that tells you that a jackass has two ears".

Quoted in John Gunther
Taken at the Flood: The Story of Albert D. Lasker (p. 96)

Mizner, Wilson
If you steal from one author, it's plagiarism; if you steal from many, it's research.

Quoted in Alva Johnston
The Legendary Mizners
Chapter 4, The Sport (p. 66)

Szent-Györgyi, Albert
Research means going out into the unknown with the hope of finding something new to bring home. If you know what you are going to do, or even to find there, then it is not research at all, then it is only a kind of honourable occupation.

Quoted in Jefferson Hane Weaver
The World of Physics
Volume II (p. 63)

von Braun, Wernher
Basic research is when I'm doing what I don't know I'm doing.

Quoted in Jefferson Hane Weaver
The World of Physics
Volume II (p. 63)

SERIES

Abel, Niels Henrik
With the exception of the geometric series, there does not exist in all of mathematics a single infinite series whose sum has been determined rigorously.

Quoted in Eli Maor
To Infinity and Beyond: A Cultural History of the Infinite (p. 29)

The divergent series are the invention of the devil, and it is a shame to base on them any demonstration whatsoever. By using them, one may draw any conclusion he pleases and that is why these series have produced so many fallacies and so many paradoxes . . .

Quoted in Eli Maor
To Infinity and Beyond: A Cultural History of the Infinite (p. 33)

Bernoulli, Jakob (Jacques)
The sum of an infinite series whose final term vanishes perhaps is infinite, perhaps finite.

Ars Conjectendi
Tractatus de Seriebus Infinitis Earumque Summa Finita et Usu
in Quadraturis Spatiorum & Ractificationibus Curvarum

Even as the finite encloses an infinite series
And in the unlimited limits appear,
So the soul of immensity dwells in minutia
And in narrowest limits no limits inhere.
What joy to discern the minute in infinity!

Ars Conjectendi
Tractatus de Seriebus Infinitis Earumque Summa Finita et Usu
in Quadraturis Spatiorum & Ractificationibus Curvarum

SET

Byron, Lord George Gordon
For a "mixt company" implies, that, save
Yourself and friends, and half a hundred more,
Whom you may bow to without looking grave,
The rest are but a vulgar set.

The Poetical Works of Lord Byron
Beppo

Cleveland, Richard
We can't be assured of a full set,
Or even a reasonable dull set.
  It wouldn't be clear
  That there's any set here,
Unless we assume there's a null set.

Mathematics Magazine
The Axioms of Set Theory
Volume 52, Number 4, September 1979 (pp. 256–7)

Poincaré, Henri
Later generations will regard *Mengenlehre* as a disease from which one
has recovered.

Apocraphyl
The Mathematical Intelligencer
Did Poincaré say "Set theory is a Disease"?
by Jeremy Gray
Volume 13, Number 1, 1991 (p. 19)

Quine, W.V.O.
We may more reasonably view set theory, and mathematics generally,
in much the way in which we view theoretical portions of the natural
sciences themselves; as comprising truths or hypotheses which are to be
vindicated less by the pure light of reason than by the indirect systematic
contribution which they make to the organizing of empirical data in the
natural sciences.

Elementary Logic (p. 4)

To say that mathematics in general has been reduced to logic hints at some new firming up of mathematics at its foundations. This is misleading. Set theory is less settled and more conjectural than the classical mathematical superstructure than can be founded upon it.

Elementary Logic (p. 125)

Reznick, Bruce
A set is a set
(you bet; you bet!)
And nothing could not be a set,
you bet!
That is, my pet
Until you've met
My very special set!
If this were a set,

It'd be a threat,
And lead to conclusions
That you'd regret
And make you fret
And wet with sweat—
This very special set!

Let A be the set of every U
That doesn't belong to U.
Then if A's in A, it's not in A
And if not, then what can you do?

So don't use the het-
Erological set
'Cause somethings cannot
Be a set, my pet.
Or better yet,
Go out and get
The *class* of every set.
Oh, Bertrand.

Mathematics Magazine
A Set is A Set
Volume 66, Number 2, April 1993 (p. 95)

SPHERE

O'Brien, Katharine
Now Einstein's Glee
was plain to see
at the sight of a cone with a sphere on top . . .

The Mathematics Teacher
Einstein and the Ice-Cream Cone
April 1968 (p. 404)

SQUARE

Unknown
A square is neither line
nor circle; it is timeless.
Points don't chase around
a square. Firm, steady,
it sits there and knows
its place. A circle
won't be squared.

Mathematical Magazine
Volume 62, Number 1, February 1989

STATISTICS

Bowley, Arthur L.
Great numbers are not counted correctly to a unit, they are estimated; and we might perhaps point to this as a division between arithmetic and statistics, that whereas arithmetic attains exactness, statistics deals with estimates, sometimes very accurate, and very often sufficiently so for their purpose, but never mathematically exact.

Elements of Statistics
Part I, Chapter I (p. 3)

A knowledge of statistics is like a knowledge of foreign languages or of algebra; it may prove of use at any time under any circumstances.

Elements of Statistics
Part I, Chapter I (p. 4)

de Madariaga, Salvador
Statistics only work well when they dwell on large numbers of absolutely free motions, or what has been described as "perfect disorder". If an element of deliberate direction, of conscious "order", meddles with their utter "innocence", the facts in question cease to follow statistical laws.

Essays with a Purpose
Freedom and Science (p. 50)

Ellis, Havelock
The methods of statistics are so variable and uncertain, so apt to be influenced by circumstance, that it is never possible to be sure that one is operating with figures of equal weight.

The Dance of Life
Chapter VII (p. 273)

Emerson, Ralph Waldo
One more fagot of these adamantine bandages is the new science of Statistics.

The Works of Ralph Waldo Emerson
Volume VI
Conduct of Life
Fate (p. 17)

Reynolds, H.T.
. . . statistics—whatever their mathematical sophistication and elegance—cannot make bad variables into good ones.

Analysis of Nominal Data (p. 8)

Stevenson, Robert Louis
Here he comes, big with statistics,
   Troubled and sharp about fac's.

The Complete Poems of Robert Louis Stevenson
LXVI

Unknown
The mark of an educated man is the ability to make a reasoned guess on the basis of insufficient information.

Source unknown

Facts are stubborn things, but statistics are more pliable.

Quoted in Evan Esar
20,000 Quips and Quotes

Statistics can be made to prove anything—even the truth.

Quoted in Evan Esar
20,000 Quips and Quotes

A statistician carefully assembles facts and figures for others who carefully misinterpret them.

Quoted in Evan Esar
20,000 Quips and Quotes

The Durbin–Whatzit statistics is used to test unknown assumptions.

Source unknown

Walker, Marshall
Mathematical statistics provides an exceptionally clear example of the relationship between mathematics and the external world. The external world provides the experimentally measured distribution curve; mathematics provides the equation (the mathematical model) that corresponds to the empirical curve. The statistician may be guided by a thought experiment in finding the corresponding equation.

The Nature of Scientific Thought (p. 50)

Wells, H.G.
Satan delights equally in statistics and in quoting scripture . . .

The Undying Fire
Chapter i, section 3

STRUCTURE

Buchanan, Scott
The structures with which mathematics deals are more like lace, the leaves of trees, and the play of light and shadow on a meadow or a human face . . .

Poetry and Mathematics (p. 36)

The structures of mathematics and the propositions about them are ways for the imagination to travel and the wings, or legs, or vehicles to take you where you want to go.

Poetry and Mathematics (p. 36)

Child, Charles Manning
Structure and function are mutually related. Function produces structure and structure modifies and determines the character of function.

Individuality in Organisms (p. 16)

SUBTRACTION

Carroll, Lewis

"Can you do Subtraction? Take nine from eight."

"Nine from eight I ca'n't, you know," Alice replied.

"She can't do Subtraction," said the White Queen.

The Complete Works of Lewis Carroll
Through the Looking Glass
Chapter IX

Try another Subtraction sum. Take a bone from a dog: what remains?

The Complete Works of Lewis Carroll
Through the Looking Glass
Chapter IX

"You don't know what you're talking about!" cried Humpty Dumpty. "How many days are there in a year?"

"Three hundred and sixty-five," said Alice.

"And how many birthdays have you?"

"One."

"And if you take one from three hundred and sixty-five what remains?"

"Three hundred and sixty-four, of course."

Humpty Dumpty looked doubtful. "I'd rather see that done on paper," he said.

The Complete Works of Lewis Carroll
Through the Looking Glass
Humpty Dumpty

Eliot, George

. . . there's no real making amends in this world, any more nor you can mend a wrong subtraction by doing your addition right.

Adam Bede
Book II, Chapter 18 (p. 291)

Gates, Bill

I think business is very simple. Profit. Loss. Take the sales, subtract the costs, you get this big positive number. The math is quite straightforward.

US News and World Report
15 February 1993

Unknown

Question: How many times can you subtract 7 from 83, and what is left afterwards?

Answer: I can subtract it as many times as I want, and it leaves 76 every time.

Source unknown

West, Mae

A man has one hundred dollars and you leave him with two dollars. That's subtraction.

The Wit and Wisdom of Mae West (p. 51)

SURFACE

Carroll, Lewis

"Now, this *third* handkerchief," Mein Herr proceeded, "has also four edges, which you can trace continuously round and round: all you need do is to join its four edges of the opening. The Purse is then complete, and its outer surface—"

"I see!" Lady Muriel eagerly interrupted. "Its *outer* surface will be continuous with its inner surface! . . . "

The Complete Works of Lewis Carroll
Sylvie and Bruno, Concluded

SYMBOLS

Buchanan, Scott
Symbols, formulae and proofs have another hypnotic effect. Because they are not immediately understood, they, like certain jokes, are suspected of holding in some sort of magic embrace the secret of the universe, or at least some of its more hidden parts.

Poetry and Mathematics (p. 36)

Each symbol used in mathematics, whether it be a diagram, a numeral, a letter, a sign, or a conventional hieroglyph, may be understood as a vehicle which someone has used on a journey of discovery.

Poetry and Mathematics (p. 47)

Hilbert, David
In the beginning there was the symbol.

Quoted in Walter R. Fuchs
Mathematics for the Modern Mind (p. 164)

Arithmetical symbols are written diagrams and geometrical figures are graphic formulas.

Bulletin of the American Mathematical Society
Mathematical Problems
Volume 8, 1902 (p. 443)

Huxley, Aldous
. . . some of the greatest advances in mathematics have been due to the invention of symbols, which it afterwards became necessary to explain; from the minus sign proceeded the whole theory of negative quantities.

Jesting Pilate
India & Burma (p. 108)

Nicholas of Cusa
If we approach the Divine through symbols, them it is most suitable that we use mathematical symbols, these have an indestructible certainty.
Quoted in Stanley Gudder
A Mathematical Journey (p. 349)

Pearson, Karl
The mathematician, carried along on his flood of symbols, dealing apparently with purely formal truths, may still reach results of endless importance for our description of the physical universe.
Quoted in Stanley Gudder
A Mathematical Journey (p. 299)

Russell, Bertrand
. . . symbolism is useful because it makes things difficult . . . Now, in the beginnings, everything is selfevident; and it is very hard to see whether one selfevident proposition follows from another or not. Obviousness is always the enemy to correctness. Hence we invent a new and difficult symbolism, in which nothing is obvious.
The International Monthly
Recent Work on the Principles of Mathematics
Volume 4, July–December 1901 (p. 85)

Truesdell, Clifford A.
There is nothing that can be said by mathematical symbols and relations which cannot also be said by words. The converse, however, is false. Much that can be and is said by words cannot successfully be put into equations, because it is nonsense.
Six Lectures on Modern Natural Philosophy
Chapter III (p. 35)

Whitehead, Alfred North
There is an old epigram which assigns the empire of the sea to the English, of the land to the French, and of the clouds to the Germans. Surely it was from the clouds that the Germans fetched + and −; the ideas which these symbols have generated are much too important to the welfare of humanity to have come from the sea or from the land.
An Introduction to Mathematics
Chapter 6 (p. 60)

SYMMETRY

Aristotle

A nose which varies from the ideal of straightness to a hook or snub may still be of good shape and agreeable to the eye.

Politics
Book V, Chapter 9, 1309b, [20]

Blake, William

Tyger, Tyger, burning bright
In the forest of the night,
What immortal hand or eye
Could frame thy fearful symmetry?

BLAKE: The Complete Poems
Songs of Innocence and of Experience
The Tyger

Borges, Jorge Luis

. . . "Reality favors symmetry".

Conversations with Jorge Luis Borges
Chapter VI (p. 109)

Carroll, Lewis

You boil it in saw dust: you salt it in glue:
You condense it with locust and tape:
Still keeping one principle object in view—
To preserve its symmetrical shape.

The Complete Works of Lewis Carroll
The Hunting of the Snark
Fit the Fifth
The Beaver's Lesson

Chesterton, Gilbert Keith

Suppose some mathematical creature from the moon were to reckon up the human body; he would at once see that the essential thing about it was that it was duplicate. A man is two men, he on the right exactly resembling him on the left. Having noted that there was an arm on the right and one on the left, a leg on the right and one on the left, he might go further and still find on each side the same number of fingers, the same number of toes, twin eyes, twin ears, twin nostrils, and even twin lobes of the brain. At last he would take it as a law; and then, where he found a heart on one side, would deduce that there was another heart on the other. And just then, where he most felt he was right, he would be wrong.

Orthodoxy
The Paradoxes of Christianity (p. 148)

Feynman, Richard P.
Why is nature so nearly symmetrical? No one has any idea why. The only thing we might suggest is something like this: There is a gate in Japan, a gate in Neiko, which is sometimes called by the Japanese the most beautiful gate in all Japan; it was built in a time when there was great influence from Chinese art. The gate is very elaborate, with lots of gables and beautiful carvings and lots of columns and dragon heads and princes carved into the pillars, and so on. But when one looks closely he sees that in the elaborate and complex design along one of the pillars, one of the small design elements is carved upside down; otherwise the thing is completely symmetrical. If one asks why this is, the story is that it was carved upside down so that the gods will not be jealous of the perfection of man. So they purposely put the error in there, so that the gods would not be jealous and get angry with human beings.

We might like to turn the idea around and think that the true explanation of the near symmetry of nature is this: that God made the laws only nearly symmetrical so that we should not be jealous of His perfection!

The Feynman Lectures on Physics
Volume 1
52-11 (p. 52-12)

Herbert, George
Man is all symmetrie,
Full of proportions, one limbe to another,
   And all to all the world besides:
   Each part may call the furthest, brother:
For head with foot hath private amitie,
   And both with moon and tides.

The Works of George Herbert
The Temple
Man

Mackay, Charles
Truth . . . and if mine eyes
Can bear its blaze, and trace its symmetries,
Measure its distance, and its advent wait,
I am no prophet—I but calculate.

The Poetical Works of Charles Mackay
The Prospects of the Future

Mao Tse-tung
Tell me why should symmetry be of importance?

30 May, 1974

Pascal, Blaise
Those who make antitheses by forcing words are like those who make false windows for symmetry.

Pensées
Section I, 27

Symmetry is what we see at a glance . . .

Pensées
Section I, 28

Updike, John

When you look	kool uoy nehW
into a mirror	rorrim a otni
it is not	ton si ti
yourself you see,	,ees uoy flesruoy
but a kind	dnik a tub
of apish error	rorre hsipa fo
posed in fearful	lufraef ni desop
symmetry	yrtemmys

Telephone Poles and Other Poems
Mirror

Valéry, Paul
The universe is built on a plan the profound symmetry of which is somehow present in the inner structure of our intellect.

Quoted in Jefferson Hane Weaver
The World of Physics
Volume II (p. 521)

Warner, Sylvia Townsend
"An umbrella, Lueli, when in use resembles the—the shell that would be formed by rotating an arc of curve about its axis of symmetry, attached to a cylinder of small radius whose axis is the same as the axis of symmetry

of the generating curve of the shell. When not in use it is properly an elongated cone, but it is more usually helicodial in form."

Lueli made no answer. He lay down again, this time face downward.

Mr. Fortune's Maggot (p. 176)

Weyl, Hermann
Symmetry, as wide or as narrow as you may define its meaning, is one idea by which man through the ages has tried to comprehend and create order, beauty, and perfection.

Symmetry (p. 5)

Symmetry is a vast subject, significant in art and nature. Mathematics lies at its root, and it would be hard to find a better one on which to demonstrate the working of the mathematical intellect.

Symmetry (p. 145)

Yang, Chen N.
Nature seems to take advantage of the simple mathematical representations of the symmetry laws. When one pauses to consider the elegance and the beautiful perfection of the mathematical reasoning involved and contrast it with the complex and far-reaching physical consequences, a deep sense of respect for the power of the symmetry laws never fails to develop.

Quoted in Heinz R. Pagels
The Cosmic Code (p. 289)

Zee, Anthony
Pick your favorite group: write down the Yang–Mills theory with your groups as its local symmetry group; assign quark fields, lepton fields, and Higgs fields to suitable representations; let the symmetry be broken spontaneously. Now watch to see what the symmetry breaks down to . . . that, essentially, is all there is to it. Anyone can play. To win, one merely has to hit on the choice used by the Greatest Player of all time. The prize? Fame and glory, plus a trip to Stockholm.

Fearful Symmetry (pp. 253–4)

TENSOR

Bell, Eric T.

. . . the tensor calculus that cost Einstein an effort to master is now a regular part of an undergraduate course in the better technical schools. The subject has been so thoroughly emulsified that even an eighteen-year-old can swallow it without regurgitating. But this does not prove that either his brain or his stomach is stronger than Einstein's was.

Mathematics: Queen and Servant of Science
A Metrical Universe (p. 211)

Benford, Gregory

There was a blithe certainty that came from first comprehending the full Einstein field equations, arabesques of Greek letters clinging tenuously to the page, a gossamer web. They seemed insubstantial when you first saw them, a string of squiggles. Yet to follow the delicate tensors as they contracted, as the superscripts paired with subscripts, collapsing mathematically into concrete classical entities—potential; mass; forces vectoring in a curved geometry—that was a sublime experience. The iron fist of the real, inside the velvet glove of airy mathematics.

Timescape
Chapter 15 (pp. 175–6)

Bester, Alfred

Tenser, said the Tensor
Tenser, said the Tensor
Tension, apprehension,
And dissension have begun.

Quoted in William L. Burke
Spacetime, Geometry, Cosmology (p. 114)

Russell, Bertrand

A man punting walks along the boat, but keeps a constant position with reference to the river bed so long as he does not pick up his pole. The Lilliputians might debate endlessly whether he is walking or standing still: the debate would be as to words, not as to facts. If we choose co-ordinates fixed relatively to the boat, he is walking; if we choose co-ordinates fixed relatively to the river bed, he is standing still. We want to express physical laws in such a way that it shall be obvious when we are expressing the same law by reference to two different systems of co-ordinates, so that we shall not be misled into supposing we have different laws when we have one law in different words. This is accomplished by the method of tensors.

The ABC of Relativity (pp. 177–8)

Unknown
The Wonderful Thing About Tensors
The wonderful thing about tensors
Are tensors have traces and norms
Their tops are made out of vectors
Their bottoms are made out of forms
There's stress and pressure
And one that measures
The distance from P to Q
But the most wonderful thing about tensors
Is the one called $G\mu\nu$

Source unknown

The decomposition of a tensor into components appears to be an artificial act, for the components of the tensor do not belong to the essential substance of the tensor. Setting up a reference system to analyze a tensor by its components is like erecting a scaffold to study a building by its parts. The scaffold does not belong to the building, but certainly fulfills a useful purpose.

Quoted in Howard W. Eves
Mathematical Circles Squared
74°

Van Dine, S.S.

The tensor is known to all advanced Mathematicians. It is one of the technical expressions used in non-Euclidean geometry; and though it was discovered by Riemann in connection with a concrete problem in physics, it has now become of widespread importance in the mathematics of relativity. It's highly scientific in the abstract sense, and so can have no direct bearing on Sprigg's murder.

The Bishop Murder Case
Chapter 9

Weyl, Hermann
The conception of tensors is possible owing to the circumstance that the transition from one co-ordinate system to another expresses itself as a **linear** transformation in the differentials. One here uses the exceedingly fruitful mathematical device of making a problem "linear" by reverting to infinitely small quantities.

Space, Time, Matter (p. 104)

Whitehead, Alfred North
The idea that physicists would in future have to study the theory of tensors created real panic amongst them following the first announcement that Einstein's predictions had been verified.

Quoted in Jean-Pierre Luminet
Black Holes (p. 47)

The mark of an educated man is the ability to make a reasoned guess on the basis of insufficient information.
Unknown – (See p. 329)

THEOREMS

Davis, Philip J.
Hersh, Reuben
If the number of theorems is larger than one can possibly survey, who can be trusted to judge what is 'important'? One cannot have survival of the fittest if there is no interaction.

The Mathematical Experience
Ulam's Dilemma (p. 21)

Rota, Gian-Carlo
Theorems are not to mathematics what successful courses are to a meal. The nutritional analogy is misleading.

Quoted in Philip J. Davis and Reuben Hersh
The Mathematical Experience
Introduction (pp. xviii–xix)

Unknown
There are no deep theorems—only theorems that we have not understood very well.

The Mathematical Intelligencer
Volume 5, Number 3, 1983

THEORIST

Cardozo, Benjamin N.
The theorist has a hard time to make his way in an ungrateful world.
He is supposed to be indifferent to realities; yet his life is spent in
the exposure of realities, which, till illuminated by his searchlight, were
hidden and unknown.

The Growth of the Law
Chapter II (p. 21)

THEORY

Cantor, Georg Ferdinand Ludwig Phillip
My theory stands as firm as a rock; every arrow directed against it will return quickly to its archer. How do I know this? Because I have studied it from all sides for many years; because I have examined all objections which have ever been made against the infinite numbers; and above all because I have followed its roots, so to speak, to the first infallible cause of all created things.

Quoted in Joseph Dauben
Georg Cantor: His Mathematics and Philosophy of the Infinite (p. 298)

Heaviside, Oliver
Theory always tends to become more abstract as it emerges successfully from the chaos of facts by processes of differentiation and elimination, whereby the essentials and their connections become recognized, while minor effects are seen to be secondary or unessential, and are ignored temporarily, to be explained by additional means.

Quoted in J.W. Mellor
Higher Mathematics for Students of Chemistry and Physics (p. 370)

Holt, Michael
Nobody knows why, but only the scientific theories that really work are the mathematical ones.

Mathematics in Art (p. 33)

Pétard, H. [Pondiczery, E.S.]

The Method of Inverse Geometry.

We place a spherical cage in the desert, enter it, and lock it. We perform an inversion with respect to the cage. The lion is then in the interior of the cage, and we are outside.

The Method of Projective Geometry.

Without loss of generality, we may regard the Sahara Desert as a plane. Project the plane into a line, and then project the line into an interior point of the cage. The lion is projected into the same point.

The "Mengentheoretisch" Method.

We observe that the desert is a separable space. It therefore contains an enumerable dense set of points, from which can be extracted a sequence having the lion as a limit. We then approach the lion stealthily along this sequence, bearing with us suitable equipment.

The Peano Method.

Construct, by standard methods, a continuous curve passing through every point of the desert. It has been shown that it is possible to traverse such a curve in an arbitrarily short time. Armed with a spear, we traverse the curve in a time shorter than that in which a lion can move his own length.

A Topological Method.

We observe that a lion has at least the connectivity of the torus. We transport the desert into four-space. It is then possible to carry out such a deformation that the lion can be returned to three-space in a knotted condition. He is then helpless.

The Cauchy, or Function Theoretical Method.

We consider an analytic lion-valued function $f(z)$. Let x be the cage. Consider the integral:

$$\frac{1}{(2\pi i)} \text{ integral over } C \text{ of } [f(z)/(z-x)] \, dz$$

where C is the boundary of the desert; its value is $f(x)$, i.e., a lion in the cage.

The Wiener Tauberian Method.

We produce a tame lion, LO of class L ($-$infinity, $+$infinity), whose Fourier transform nowhere vanishes, and release it in the desert. LO then converges to our cage. By Wiener's General Tauberian Theorem, any other lion, L (say), will then converge to the same cage. Alternatively, we can approximate arbitrarily close to L by translating LO about the desert.

The American Mathematical Monthly
A Contribution to the Mathematical Theory of Big Game Hunting
Volume 45, 1938

Truesdell, Clifford

The hard facts of classical mechanics taught to undergraduates today are, in their present forms, creations of James and John Bernoulli, Euler, Lagrange, and Cauchy, men who never touched a piece of apparatus; their only researches that have been discarded and forgotten are those where they tried to fit theory to experimental data. They did not disregard experiment; the parts of their work that are immortal lie in domains where experience, experimental or more common, was at hand, already partly understood through various special theories, and they abstracted and organized it and them. To warn scientists today not to disregard experiment is like preaching against atheism in church or communism among congressmen. It is cheap rabble-rousing. The danger is all the other way. Such a mass of experimental data on everything pours out of organized research that the young theorist needs some insulation against its disrupting, disorganizing effect. Poincaré said, "The scientist must order; science is made out of facts as a house is made out of stones, but an accumulation of facts is no more science than a heap of stones, a house." Today the houses are buried under an avalanche of rock splinters, and what is called theory is often no more than the trace of some moving fissure on the engulfing wave of rubble. Even in earlier times there are examples. Stokes derived from his theory of fluid friction the formula for the discharge from a circular pipe. Today this classic formula is called the "Hagen–Poiseuille law" because Stokes, after comparing it with measured data and finding it did not fit, withheld publication. The data he had seem to have concerned turbulent flow, and while some experiments that confirm his mathematical discovery had been performed, he did not know of them.

Six Lectures on Modern Natural Philosophy
Chapter VI (pp. 92–3)

THEORY OF FUNCTIONS

Keyser, Cassius J.
The Modern Theory of Functions—that stateliest of all the pure creations of the human intellect.

Quoted in Columbia University
Lectures on Science, Philosophy and Art 1907–1908 (p. 16)

Volterra, Vito
The theory that has had the greatest development in recent times is without any doubt the theory of functions.

Quoted in Stanley Gudder
A Mathematical Journey (p. 32)

THOUGHT

Whitehead, Alfred North
. . . the first man who noticed the analogy between a group of seven
fishes and a group of seven days made a notable advance in the history
of thought.

Science and the Modern World
Mathematics as an Element in the History of Thought (p. 30)

TOPOLOGIST

Unknown

A topologist is someone who can't tell the difference between his ass and a hole in the ground, but who can tell the difference between his ass and two holes in the ground.

Source unknown

In the beginning there was only one kind of Mathematician, created by the Great Mathematical Spirit from the Book: the Topologist. And they grew to large numbers and prospered.

One day they looked up in the heavens and desired to reach up as far as the eye could see. So they set out in building a Mathematical edifice that was to reach up as far as "up" went. Further and further up they went . . . until one night the edifice collapsed under the weight of paradox.

The following morning saw only rubble where there once was a huge structure reaching to the heavens. One by one, the Mathematicians climbed out from under the rubble. It was a miracle that nobody was killed; but when they began to speak to one another, SURPRISE of all surprises! they could not understand each other. They all spoke different languages. They all fought amongst themselves and each went about their own way. To this day the Topologists remain the original Mathematicians.

Source unknown

TOPOLOGY

Unknown

You could define the subspace topology this way, if you were sufficiently malicious.

Mathematical and Scientific Quotations from Cambridge
The Internet

I THINK SIR WILL BE IMPRESSED
BY THE CONTENTS OF THIS DRAWER..!

Logic is a large drawer . . . A wise man will look into it for two purposes, to avail himself of those instruments that are really useful, and to admire the ingenuity with which those that are not so, are assorted and arranged.

Charles Caleb Colton – (See p.119)

TRANSCENDENTAL NUMBERS

Sagan, Carl
Hiding between all the ordinary numbers was an infinity of transcendental numbers whose presence you would never have guessed until you looked deeply into mathematics.

Contact: A Novel (p. 21)

TRANSFORM

Unknown

A polar bear is a rectangular bear after a coordinate transform.

Source unknown

YOU KNOW THAT PEBBLE YOU JUST THREW...?!

It is a mathematical fact that the casting of this pebble from my hand alters the centre of gravity of the universe.
Thomas Carlyle – (See p.126)

TRANSITIONS

Unknown

Just because they are called 'forbidden' transitions does not mean that they are forbidden. They are less allowed than allowed transitions, if you see what I mean.

Mathematical and Scientific Quotations from Cambridge
The Internet

TRIANGLE

Beckett, Samuel

. . . and do not despair. Remember there is no triangle, however obtuse, but the circumference of some circle passes through its wretched vertices.

Murphy
Chapter 10 (p. 213)

Mathematicians, whose tempers are generally intolerable, are perhaps psychologically excusable, for the constant tension of their mind is, perhaps, the cause of their bad digestion and their state of hypochondria.

Camille Flammarion – (See p. 157)

TRIGONOMETRY

Chesterton, Gilbert Keith
A straight liner is straight
And a square mile is flat:
But you learn in trigonometrics a trick worth two of that.

The Collected Poems of G.K. Chesterton
Songs of Education
V. The Higher Mathematics (p. 97)

Howell, Scott, Sen. [D-Sandy]
As long as schools continue to teach trigonometry and algebra, there will always be a moment of silence, and indeed prayer, in our public schools.

On why he sees no need to formalize a moment of silence in Utah schools

Unknown
Trigonometry is a sine of the times!

Source unknown

TRUTH

d'Alembert, Jean le Rond
Geometrical truths are in a way asymptotes to physical truths, that is to say, the latter approach the former indefinitely neat without ever reaching them exactly.

Quoted in Alphonse Rebière
Mathématiques et Mathématiciens: Pensées et Curiosités (p. 10)

Duhem, Pierre
Unlike the reduction to absurdity employed by geometers, experimental contradiction does not have the power to transform a physical hypothesis into an indisputable truth; in order to confer this power on it, it would be necessary to enumerate completely the various hypotheses which may cover a determinate group of phenomena; but the physicist is never sure he has exhausted all the imaginable assumptions. The truth of a physical theory is not decided by heads or tails.

The Aim and Structure of Physical Theory (p. 190)

Everett, Edward
In the pure mathematics we contemplate absolute truths, which existed in the Divine Mind before the morning stars sang together, and which will continue to exist there, when the last of their radiant host shall have fallen from heaven.

Quoted in E.T. Bell
Mathematics: Queen and Servant of Science
Mathematical Truth (p. 21)

Huxley, Aldous
All great truths are obvious truths. But not all obvious truths are great truths.

Music at Night and Other Essays (p. 17)

Jeffers, Robinson
The mathematicians and physics men
Have their mythology; they work alongside the truth,
Never touching it; their equations are false
But the things *work*. Or, when gross error appears,
They invent new ones; they drop the theory of waves
In universal ether and imagine curved space.

The Beginning and the End
The Great Wound

Tolstoy, Leo
Some mathematician, I believe, has said that true pleasure lies not in the discovery of truth, but in the search for it.

Anna Karenina
Part II
Chapter XIV (p. 192)

Unknown
The Cartesian criterion of truth.

Source unknown

Wilde, Oscar
It is a terrible thing for a man to find out suddenly that all his life he has been speaking nothing but the truth.

Quoted in John D. Barrow
The World within the World (p. 260)

JACK— . . . That, my dear Algy, is the whole truth, pure and simple.

ALGERNON—The truth is rarely pure and never simple.

The Importance of Being Earnest
Act I

Wilkins, John
That the strangeness of this opinion is no sufficient reason why it should be rejected, because other certain truths have been formerly esteemed ridiculous, and great absurdities entertayned by common consent.

The Discovery of a World in the Moone (p. 1)

VECTOR

Thomson, William [Lord Kelvin]
Quaternions came from Hamilton . . . and have been an unmixed evil to those who have touched them in any way. Vector is a useless survival . . . and has never been of the slightest use to any creature.

Quoted in Jerrold E. Marsden and Anthony J. Tromba
Vector Calculus (p. 1)

WISDOM

Bell, Eric T.
Wisdom was not born with us, nor will it perish when we descend into the shadows with a regretful backward glance that other eyes than ours are already lit by the dawn of a new and truer mathematics.

The Queen of the Sciences
Chapter X (p. 138)

ZERO

Bôcher, Maxime

. . . there is what may perhaps be called the method of optimism which leads us either willfully or instinctively to shut our eyes to the possibility of evil. Thus the optimist who treats a problem in algebra or analytic geometry will say, if he stops to reflect on what he is doing: "I know that I have no right to divide by zero; but there are so many other values which the expression by which I am dividing might have that I will assume that the Evil One has not thrown a zero in my denominator this time."

Bulletin of the American Mathematical Society
2nd Series
The Fundamental Conceptions and Methods in Mathematics
Volume 11, 1904 (pp. 134–5)

Hugo, Victor

One microscopic glittering point; then another; and another, and still another; they are scarcely perceptible, yet they are enormous. This light is a focus; this focus, a star; this star, a sun; this sun, a universe; this universe, nothing. Every number is zero in the presence of the infinite.

The Toilers of the Sea
Volume II (p. 124)
Book II, Labor
Sub Umbra

Unknown

Sometimes it is useful to know how large your zero is.

Source unknown

Zero is more than nothing!

Source unknown

Thank you for not dividing by zero.

Source unknown

Zamyatin, Yevgeny

The circles differ—some are golden, some bloody. But all are equally divided into three hundred and sixty degrees. And the movement is from zero—onward, to ten, twenty, two hundred, three hundred and sixty degrees—back to zero. Yes, we have returned to zero— yes. But to my mathematical mind it is clear that this zero is altogether different, altogether new. We started from zero to the right, we have returned to it from the left. Hence, instead of plus zero, we have minus zero. Do you understand?

We
Twentieth Entry (p. 103)

Sometimes it is useful to know how large your zero is.
Unknown – (See p. 361)

BIBLIOGRAPHY

Abbott, Edwin A. *Flatland*. Barnes & Noble, Inc., New York. 1963.

Adams, Douglas. *Mostly Harmless*. Harmony Books, New York. 1992.

Adams, Douglas. *The Original Hitchhiker Radio Scripts*. Crown Publishers, Inc., New York. 1985.

Adams, Franklin. *Tobogganing on Parnassus*. Doubleday, Page I Company, Garden City. 1918.

Adams, Henry. *A Letter to American Teachers of History*. S.H. Furs & Co., Baltimore. 1910.

Adams, Henry. *The Education of Henry Adams*. Random House, New York. 1946.

Adler, Alfred. 'Mathematics and Creativity' in *New Yorker Magazine*. February 19, 1972.

Adler, Irving. *A New Look At Geometry*. The John Day Company, New York. 1966.

Aeschylus. *The Plays of Aeschylus* in *Great Books of the Western World*. Translated by G.M. Cookson. Volume 5. Encyclopædia Britannica, Inc., Chicago. 1952.

Albers, Donald J., Alexanderson, Gerald L. and Reid, Constance. *More Mathematical People*. Harcourt Brace Jovanovich, Publishers, Boston. 1990.

Allen, Arnold O. *Probability, Statistics, and Queuing Theory with Computer Science Applications* (Second Edition). Academic Press, Inc., Boston. 1990.

Allen, William. 'On the Curves of Trisection' in *American Journal of Science*. Volume 4. 1822.

Allen, Woody. *Getting Even*. Random House, New York. 1971.

Anderson, G.L. 'An Interview with Constance Reid' in *Two Year College Mathematical Journal*. Volume 11. 1980.

Andrews, William Symes. *Magic Squares and Cubes*. Dover Publications, New York. 1960.

Anglin, W.S. 'Mathematics and History' in *Mathematical Intelligencer*. Volume 14, Number 4. Fall 1992.

Anscombe, F.J. 'Rejection of Outliers' in *Technometrics*. Volume 2. 1960.

363

Apostle, Hippocrates George. *Aristotle's Philosophy of Mathematics*. The University of Chicago Press, Chicago. 1952.

Arago. 'Eulogy on Laplace' in *Smithsonian Report*. 1874.

Arbuthnot, John. *Of the Laws of Chance*. Benj. Matte, London. 1692.

Archibald, R.C. 'Mathematicians and Music' in *The American Mathematical Monthly*. Volume 31. January 1924.

Aristotle. *Generation of Animals* in *Great Books of the Western World*. Translated by R.P. Hardie and R.K. Gaye. Volume 8. Encyclopædia Britannica, Inc., Chicago. 1952.

Aristotle. *Metaphysica* in *Great Books of the Western World*. Translated by E.W. Webster. Volume 8. Encyclopædia Britannica, Inc., Chicago. 1952.

Aristotle. *On Generation and Corruption* in *Great Books of the Western World*. Translated by R.P. Hardie and R.K. Gaye. Volume 8. Encyclopædia Britannica, Inc., Chicago. 1952.

Aristotle. *Physics* in *Great Books of the Western World*. Translated by R.P. Hardie and R.K. Gaye. Volume 8. Encyclopædia Britannica, Inc., Chicago. 1952.

Aristotle. *Politics* in *Great Books of the Western World*. Translated by Benjamin Jowett. Volume 9. Encyclopædia Britannica, Inc., Chicago. 1952.

Aristotle. *Posterior Analytics* in *Great Books of the Western World*. Translated by G.R.G. Mure. Volume 8. Encyclopædia Britannica, Inc., Chicago. 1952.

Arnauld, Antoine. *The Art of Thinking: Port-Royal Logic*. The Bobbs-Merrill Co., Indianapolis. 1964.

Asimov, Isaac. *Foundation and Earth*. Doubleday, Garden City. 1986.

Asimov, Isaac. *Prelude to Foundation*. Doubleday, New York. 1988.

Aurelius, Marcus. *Meditations of the Emperor Marcus Aurelius Antonius*. Translated by R. Graves. W. Baynes, London. 1811.

Auster, Paul. *The Music of Chance*. Viking, New York. 1990.

Babbage, Charles. *Charles Babbage and His Calculating Engines*. Dover Publications, Inc., New York. 1961.

Bacon, Francis. *The Advancement of Learning* in *Great Books of the Western World*. Volume 30. Encyclopædia Britannica, Inc., Chicago. 1952.

Bacon, Francis. *Bacon's Essays*. The Macmillan Company, New York. 1930.

Bacon, Francis. *Novum Organum* in *Great Books of the Western World*. Volume 30. Encyclopædia Britannica, Inc., Chicago. 1952.

Bacon, Roger. *The Opus Majus of Roger Bacon*. Volume 1. A translation by Robert Belle Burke. Russell & Russell, Inc. New York. 1962.

Baez, Joan. *Daybreak*. Avon Books, New York. 1968.

Bain, Alexander. *Education as a Science*. D. Appleton and Co., New York. 1897.

Baker, W.R. 'The Magic Box' in *Harper's Magazine*. Volume CLXI. 1928.

Baring-Gould, William S. *The Lure of the Limerick*. Clarkson, N. Potter, Inc., New York. 1967.

Barnett, P.A. *Common Sense in Education and Teaching*. Longmans, Green, and Co., London. 1899.

Barrie, James Matthew. *The Plays of J.M. Barrie*. Charles Scribner's Sons, New York. 1948.

Bartlett, Albert A. 'The Exponential Function' in *The Physics Teacher*. Volume 14, Number 7. October 1976.

Barrow, John D. *Pi in the Sky*. Clarendon Press, Oxford. 1992.

Barrow, John D. *The Artful Universe*. Clarendon Press, Oxford. 1995.

Barrow, John D. *The World within the World*. Clarendon Press, Oxford. 1988.

Barrow, John D. *Theories of Everything*. Clarendon Press, Oxford. 1991.

Barry, Frederick. *The Scientific Habit of Thought*. Columbia University Press, New York. 1927.

Bateson, Gregory. *Steps to an Ecology of Mind*. Ballantine Books, New York. 1972.

Baumel, Judith. *The Weight of Numbers*. Wesleyan University Press, Middletown. 1988.

Bean, William B. *Aphorisms from Latham*. Prarie Press, Iowa City. 1962.

Beard, George M. 'Experiments with Living Human Beings' in *Popular Science Monthly*. Volume 14. April 1879.

Beckett, Samuel. *Murphy*. Grover Press, Inc., New York. 1957.

Beckmann, Petr. *A History of Pi*. The Golem Press, Boulder. 1982.

Beerbohm, Max. *Mainly on the Air*. Alfred A. Knopf, New York. 1958.

Begley, Sharon. 'New Answers for an Old Question' in *Newsweek*. July 5, 1993.

Bell, E.T. *Debunking Science*. University of Washington Book Store, Seattle. 1930.

Bell, Eric T. *Mathematics: Queen and Servant of Science*. McGraw-Hill Book Co., Inc., New York. 1940.

Bell, Eric T. *Men of Mathematics*. Simon and Schuster, New York. 1937.

Bell, Eric T. *The Development of Mathematics*. McGraw-Hill Book Company, Inc., New York. 1940.

Bell, Eric T. *The Handmaiden of Sciences*. The Williams & Wilkins Company, Baltimore. 1937.

Bell, Eric T. *The Magic of Numbers*. McGraw-Hill Book Company, Inc., New York. 1946.

Bell, Eric T. *The Queen of the Sciences*. The Williams & Wilkins Company, Baltimore. 1931.

Bell, Eric T. *The Search for Truth*. George Allen & Unwin Ltd., London. 1949.

Bellman, Richard. *Eye of the Hurricane: An Autobiography*. World Scientific, Singapore. 1984.

Bellman, Richard Ernest. *A Brief Introduction to Theta Functions*. Holt, Rinehart and Winston, New York. 1961.

Belloc, Hilaire. *The Aftermath*. Duckworth, London. 1927.

Benford, Gregory. *Timescape*. Simon and Schuster, New York. 1980.

Bennett, Charles H. *Old Nurse's Book of Rhymes, Jingles and Ditties*. Griffith and Farran, London. 1858.

Bentley, Arthur F. *Linguistic Analysis of Mathematics*. The Principia Press, Inc., Bloomington. 1932.

Bergson, Henri. *Creative Evolution*. Henry Holt and Company, New York. 1911.

Berkeley, Edmund C. 'Right Answers—A Short Guide for Obtaining Them' in *Computers and Automation*. September 1969.

Berkeley, George. *The Analyst*. Edited and translated by Douglas M. Jesseph. Kluwer Academic Publishers, Dordrecht. 1992.

Berkeley, George. *The Principles of Human Knowledge*. P. Smith, Glouster. 1978.

Berrett, Wayne. 'It Had to be e' in *Mathematics Magazine*. Volume 68, Number 1. February 1995.

Beveridge, W.I.B. *The Art of Scientific Investigation*. Norton, New York. 1957.

Billings, Josh. *Old Probability: Perhaps Rain—Perhaps Not*. Literature House, Upper Saddle River. 1970.

Birns, Harold. 'City Acts to Unify Inspection Rules' in *New York Times*. October 2, 1963.

Bishop, Errett. *Foundations of Constructive Analysis*. McGraw-Hill Book Company, New York. 1967.

Black, Max. *The Nature of Mathematics*. London. 1933.

Blake, William. *BLAKE: The Complete Poems*. Longman, London. 1989.

Boas, Ralph P., Jr. *Lion Hunting & Other Mathematical Pursuits*. Mathematical Association of America, Washington, D.C. 1995.

Bôcher, Maxime. 'The Fundamental Conceptions and Methods in Mathematics' in *Bulletin of the American Mathematical Society*. Volume 11, 2nd series. 1904–1905.

Bochner, Salomon. *The Role of Mathematics in the Rise of Science*. Princeton, New Jersey. 1966.

Boehm, George A. *The New World of Mathematics*. Faber and Faber, London. 1959.

Boole, George. 'An Investigation of the Laws of Thought' in *Collected Logical Works*. Volume II. The Open Court Publishing Co., LaSalle. 1952.

Borel, Émile. *Géométrie, premier et second cycles*. Ed. 2. Paris. 1908.

Borel, Émile. *Probability and Certainty*. Walker, New York. 1963.

Borel, Émile. *Probabilities and Life*. Translated by Maurice Baudin. Dover Publications, New York. 1962.

Borges, Jorge Luis. *Conversations with Jorge Luis Borges.* Holt, Rinehart and Winston, New York. 1968.

Borges, Jorge Luis. *Ficciones.* Grove Press, New York. 1962.

Born, Max. *Einstein's Theory of Relativity.* Dover Publications, Inc., New York. 1965.

Boswell, James. *The Life of Samuel Johnson.* International Collectors Library, Garden City. 1945.

Bourbaki, Nicolas. 'Nicolas Bourbaki' in *Scientific American.* May 1957.

Boutroux, E. *Natural Law in Science and Philosophy.* D. Nutt, London. 1914.

Bowley, Arthur L. *Elements of Statistics.* Staple Press Limited, London. 1946.

Box, G.E.P. 'Discussion' in *Journal of the Royal Statistical Society.* Ser. B. Volume 18, 1956.

Boyd, T.A. *Professional Amateur.* E.P. Dutton & Co., Inc., New York. 1957.

Boyd, William. *Brazzaville Beach.* Sinclair-Stevenson Limited, London. 1990.

Boyer, Carl B. *A History of Mathematics.* Wiley, New York. 1968.

Boyer, Carl B. *The History of the Calculus and its Conceptual Development.* Dover Publications, Inc., New York. 1949.

Boyer, Carl B. 'The Invention of Analytic Geometry' in *Scientific American.* Volume 180, Number 1. January 1945.

Brewster, G.W. 'Mathematical Notes' in *The Mathematical Gazette.* Volume 25, Number 263. February 1941.

Bridges, Robert. *The Testament of Beauty.* Oxford University Press, New York. 1930.

Bridgman, P.W. *The Logic of Modern Physics.* The Macmillan Company, New York. 1961.

Bridgman, Percy William. *The Way Things Are.* Harvard University Press, Cambridge. 1959.

Bronowski, Jacob. *The Ascent of Man.* Little, Brown and Company, Boston. 1973.

Brown, Fredric. *And the Gods Laughed.* Phantasia Press, West Bloomfield. 1987.

Browne, Thomas. *Hydriotaphia.* Arno Press, New York. 1977.

Browne, Thomas. *Religio Medici.* Clarendon Press, Oxford. 1909.

Browne, Thomas. *The Garden of Cyrus.* University Press, Cambridge. 1958.

Brownell, William A. 'The Revolution in Arithmetic' in *The Arithmetic Teacher.* Volume 1, Number 1. February 1954.

Browning, Elizabeth Barrett. *The Complete Poetical Works of Elizabeth Barrett Browning.* Houghton, Mifflin and Company, Boston. 1900.

Bruner, Jerome Seymore. *On Knowing—Essays for the Left Hand.* Harvard University Press, Cambridge. 1979.

Bruner, Jerome Seymore. *The Process of Education.* Harvard University Press, Cambridge. 1960.

Buchanan, Scott. *Poetry and Mathematics*. J.B. Lippincott Company, Philadelphia. 1962.

Büchner, Ludwig. *Force and Matter*. Truth Seeker Co., New York. 1950.

Bullock, James. 'Literacy in the Language of Mathematics' in *The American Mathematical Monthly*. Volume 101, Number 8. October 1994.

Bunch, Bryan H. *Mathematical Fallacies and Paradoxes*. Van Nostrand Reinhold Company, New York. 1982.

Burchfield, Joe D. *Lord Kelvin and the Age of the Earth*. Science History Publications, New York. 1974.

Burke, Edmund. *On the Sublime and Beautiful*. Printed for D. Johnson, Portland by J. Watts, Philadelphia. 1806.

Burke, Edmund. *Reflections on the French Revolution*. Hackett Publishing Company, Indianapolis. 1987.

Burke, William L. *Spacetime, Geometry, Cosmology*. University Science Books, Mill Valley. 1980.

Burr, E. Scott. 'A Rhyming Π' in *The Physics Teacher*. February 1979.

Burton, David M. *The History of Mathematics*. Wm. C. Brown Publishers, Dubuque. 1988.

Bush, Vannevar. *Endless Horizons*. Public Affairs Press, Washington, D.C. 1946.

Butler, Nicholas Murray. *The Meaning of Education, and Other Essays and Addresses*. The Macmillan Company, New York. 1898..

Butler, Samuel. *The Poetical Works of Samuel Butler*. Volume I. J. Nichol, Edinburgh. 1854.

Byron, Lord. *The Poetical Works of Lord Byron*. Jas. B. Smith & Company, Philadelphia. 1859.

Cajori, Florian. *A History of Mathematics*. Chelsea Publishing Co., New York. 1980.

Cajori, Florian. *The Teaching and History of Mathematics in the United States*. Government Printing Office, Washington D.C. 1890.

Campbell, Douglas M. and Higgins, John C. *Mathematics: People, Problems, Results*. Volume III. Wadsworth International, Belmont. 1984.

Camus, Albert. *The Fall*. Vintage Books, New York. 1956.

Cardozo, Benjamin N. *The Growth of the Law*. Yale University Press, New Haven. 1924.

Carlyle, Thomas. *English and Other Critical Essays*. J.M. Dent & Sons Ltd., London. 1950.

Carlyle, Thomas. *Sartor Resartus*. G. Bell & Sons, London. 1909.

Carmichael, R.D. *The Logic of Discovery*. The Open Court Publishing Co., Chicago. 1930.

Carroll, Lewis. *The Complete Works of Lewis Carroll*. The Modern Library, New York. 1936.

Carus, Paul. 'Logical and Mathematical Thought' in *The Monist*. Volume XX, Number 1. January 1910.

Carus, Paul. 'The God Problem' in *The Monist*. Volume XVI, Number 1. January 1906.

Casson, Stanley. *Progress and Catastrophe*. Hamish Hamilton, London. 1937.

Casti, John L. *Reality Rules*. Volume I. John Wiley & Sons, Inc., New York. 1992.

Cedering, Siv. *Letters from the Floating World*. University of Pittsburg Press, Pittsburg. 1984.

Chandrasekharan, K. *Hermann Weyl*. Springer-Verlag, Berlin. 1986.

Chapman, C.H. 'The Theory of Transformation Groups' in *New York Mathematical Society Bulletin*. Volume 2, Number 14 (First Series). 1893.

Chekhov, Anton. *Note-Book of Anton Chekhov*. Translated by S.S. Koteliansku and Leonard Woolf. B.W. Huebsch, Inc., New York. 1922.

Chesterton, G.K. *Lunacy and Letters*. Sneed & Ward, New York. 1958.

Chesterton, G.K. *Orthodoxy*. Garden City Publishing Company, Inc., Garden City. 1908.

Chesterton, G.K. *The Collected Poems of G.K. Chesterton*. Dodd, Mead & Company, New York. 1932.

Chesterton, G.K. *The G.K. Chesterton Calendar*. Cecil Palmer & Hayward, London. 1916.

Child, J.M. *The Early Mathematical Manuscripts of Leibniz*. The Open Court Publishing Company, Chicago. 1920.

Child, M.C. *Individuality in Organisms*. The University of Chicago Press, Chicago. 1915.

Chrystal, G. *Algebra*. Part II. Sixth Edition. Chelsea Publishing Company, New York. 1959.

Churchill, Winston S. *Lord Randolph Churchill*. Volume II. The Macmillan Company, London. 1906.

Churchill, Winston S. *My Early Life: A Roving Commission*. Charles Scribner's Sons, New York. 1958.

Clark, Ronald W. *Einstein: The Life and Times*. World Publishers, New York. 1971.

Cleveland, Richard. 'The Axioms of Set Theory' in *Mathematics Magazine*. Volume 52, Number 4. September 1979.

Clifford, William Kingston. *Common Sense in the Exact Sciences*. A.A. Knopf, New York. 1946.

Cochran, William G. and Cox, Gertrude M. *Experimental Designs*. John Wiley & Sons, Inc., New York. 1957.

Cohen, John. *Chance, Skill, and Luck*. Penguin Books, Baltimore. 1960.

Cohen, Morris. *A Preface to Logic*. H. Holt and Company, New York. 1944.

Cohen, R.S., Stachel, J.J. and Wartofsky, M.W. *Boston Studies in the Philosophy of Science*. Volume XV. D. Reidel Publishing Company, Dordrecht. 1974.

Coleridge, Samuel Taylor. *The Complete Poetical Works of Samuel Taylor Coleridge*. Volume I. The Clarendon Press, Oxford. 1912.

Coleridge, Samuel Taylor. *Lectures and Notes on Shakespeare and Other English Poets*. George Bell and Sons, London. 1908.

Coleridge, Samuel Taylor. *The Theory of Life*. George Bell & Sons, New York. 1892.

Colton, Charles C. *Lacon*. William Gowans, New York. 1849.

Columbia University. *Lectures On Science, Philosophy and Art 1907–1908*. The Columbia University Press, New York. 1908.

Comte, Auguste. *The Positive Philosophy*. Volume I. John Chapman, London. 1875.

Comte, Auguste. *Philosophy of Mathematics*. Harper & Brothers, Publishers, New York. 1851.

Conant, James B. *Science and Common Sense*. Franklin Watts, Inc., Publishers, New York. 1951.

Conrad, Joseph. *The Secret Agent*. Doubleday, Page & Company, Garden City. 1917.

Cooley, Hollis R., David Gans, Morris Kline and Howard E. Wahlert. *Introduction to Mathematics*. Houghton Mifflin Company, Boston. 1937.

Coolidge, Julian Lowell. *A Treatise on Algebraic Plane Curves*. Dover Publications, Inc., New York. 1959.

Copernicus, Nicolaus. *On the Revolutions of the Heavenly Spheres* in *Great Books of the Western World*. Translated by Charles Glenn Wallis. Volume 16. Encyclopædia Britannica, Inc., Chicago. 1952.

Cort, David. *Social Astonishments*. The Macmillan Company, New York. 1963.

Courant, Richard. 'Mathematics in the Modern World' in *Scientific American*. Volume 211, Number 3. September 1964.

Courant, R. and Hilbert, D. *Methods of Mathematical Physics*. Volume I. Interscience, New York. 1953.

Courant, Richard and Robbins, Herbert. *What is Mathematics?* Oxford University Press, London. 1941.

Cournot, Augustin. *Researches into the Mathematical Principles of the Theory of Wealth*. The Macmillan Company, New York. 1927.

Court, Nathan A. *Mathematics in Fun and in Earnest*. The Dial Press, New York. 1958.

Cousins, Norman. 'Editor's Odyssey' in *Saturday Review*. April 15, 1978.

Crane, Hart. *The Collected Poems of Hart Crane*. Liverlight Publishing Corporation, New York. 1946.

Crawford, Franzo H. *Introduction to the Science of Physics*. Harcourt, Brace & World, Inc., New York. 1968.

Crew, Henry. *General Physics*. The Macmillan Company, New York. 1927.

Crichton, Michael. *Rising Sun*. Alfred A. Knopf, New York. 1992.

Crick, Francis. *What Mad Pursuit*. Basic Books, Inc., Publishers, New York. 1988.

Cross, Hardy. *Engineers and Ivory Towers*. McGraw-Hill Book Company, Inc., New York. 1952.

Czarnomski, F.B. *The Wisdom of Winston Churchill*. George Allen and Unwin Ltd., London. 1956.

d'Abro, A. *The Evolution of Scientific Thought from Newton to Einstein*. Dover Publications, Inc., New York. 1950.

d'Abro, A. *The Decline of Mechanism*. D. Van Nostrand Company, Inc., New York. 1939.

da Vinci, Leonardo. *The Notebooks of Leonardo da Vinci*. Volumes I and II. Translated by Edward MacCurdy. Reynal & Hitchcock, New York. 1938.

Dantzig, Tobias. *Number: The Language of Science*. The Macmillan Company, New York. 1954.

Darwin, Charles. *The Life and Letters of Charles Darwin*. Volume I. D. Appleton and Company, New York. 1888.

Dauben, Joseph. *George Cantor: His Mathematics and Philosophy of the Infinite*. Harvard University Press, Cambridge. 1979.

Davies, Paul. *About Time*. Simon and Schuster, New York. 1995.

Davies, Paul. *Superforce*. Simon and Schuster, New York. 1984.

Davies, Robertson. *The Table Talk of Samuel Marchbanks*. Clarke, Irwin & Company Limited, Toronto. 1949.

Davis, Philip J. 'Numbers' in *Scientific American*. Volume 211, Number 3. September 1964.

Davis, Philip J. and Hersh, Reuben. *The Mathematical Experience*. Birkhauser, Boston. 1981.

Day-Lewis, C. *Collected Poems 1929–1933*. Random House, New York. 1935.

de Bruijn, N.G. *Asymptotic Methods in Analysis*. North Holland Publishing Company, Amsterdam. 1958.

de Fontenelle, Bernard. *Histoire du Renouvellement de l'Académie des Sciences*. Culture et Civilisation, Bruxelles, 1969.

de Jouvenel, Bertrand. *The Art of Conjecture*. Basic Books, New York. 1967.

de Madariaga, Salvador. *Essays with a Purpose*. Hollis & Carter, London. 1954.

de Morgan, Augustus. *A Budget of Paradoxes*. Volumes I and II. The Open Court Publishing Co., Chicago. 1915.

de Morgan, Augustus. *On the Study and Difficulties of Mathematics*. The Open Court Publishing Company, La Salle. 1943.

de Morgan, Augustus. *The Differential and Integral Calculus*. Published under the superintendence of the Society for the Diffusion of Useful Knowledge. R. Baldwin, London. 1842.

de Morgan, Augustus. *Transactions of the Cambridge Philosophical Society*. Volume 8, 1844.

de Saint-Exupéry, Antoine. *The Little Prince*. Translated by Katherine Woods. Harcourt Brace Jovanovich, Publishers, New York. 1971.

Dehn, Max. 'The Mentality of the Mathematician' in *The Mathematical Intelligencer*. Volume 5, Number 2. 1983.

Descartes, René. 'Rules for the Direction of the Mind' in *Great Books of the Western World*. Translated by Elizabeth S. Haldane and G.R.T. Ross. Volume 31. Encyclopædia Britannica, Inc., Chicago. 1952.

Dewey, John. *Democracy and Education*. The Macmillan Company, New York. 1916.

DeWitt, Cecile M. and Wheeler, John A. *Battelle Rencontres*. W.A. Benjamin, Inc., New York. 1968.

Diamond, Solomon. *Information and Error*. Basic Books, Inc., New York. 1959.

Dickens, Charles. *Martin Chuzzlewit*. T.B. Peterson & Brothers, Philadelphia. 1870.

Dickson, L.E. *History of the Theory of Numbers*. Chelsea Publications Co., New York. 1966.

Dickson, Paul. *The Official Rules*. Delcorte, Press, New York. 1978.

Dieudonné, J.A. 'The Works of Nicholas Bourbaki' in *The American Mathematical Monthly*. Volume 77. 1970.

Dieudonné, Jean: *Mathematics—The Music of Reason*. Springer-Verlag, Berlin. 1987.

Dirac, Paul Adrien Maurice. 'Methods in Theoretical Physics' in *From a Life of Physics*. Evening lectures at the International Center for Theoretical Physics, Trieste, Italy. No date.

Dirac, Paul Adrien Maurice. 'Quantised Singularities in the Electromagnetic Field' in *Proceedings of the Royal Society*. Series A, Volume 133, Number 821. 1931.

Dirac, Paul Adrien Maurice. 'The Evolution of the Physicist's Picture of Nature' in *Scientific American*. Volume 208, Number 5. May 1963.

Dodgson, Charles L. *A New Theory of Parallels*. Macmillan and Co., London. 1890.

Dodgson, C.L. 'Freeman Dyson: Mathematician, Physicist, and Writer' in *The College Mathematics Journal*. Volume 25, Number 1. January 1994.

Dostoevsky, Fyodor Mikhailovich. *The Brothers Karamozov*. Macmillan, New York. 1955.

Dostoevsky, Fyodor Mikhailovich. *The Short Novels of Dostoevsky*. Dial Press, New York. 1951.

Doyle, Arthur Conan. *The Complete Sherlock Holmes*. Garden City Publishing Co., New York. 1930.

Dryden, John. *The Poetical Works of Dryden*. Macmillan, New York. 1904.

Dudley, Underwood. 'Formulas for Primes' in *Mathematical Magazine* Volume 56. 1983.

Duffin, R.J. 'The Patron Saint of Mathematics' in *The Mathematica Intelligencer*. Volume 15, Number 1. 1993.

Duhem, Pierre. *The Aim and Structure of Physical Theory*. Atheneum, New York. 1977.

Dunsany, Lord Edward John Moreton Drax Plunkett. *My Ireland*. Jarrold Publishers, London. 1937.

Duren, Peter. *A Century of Mathematics in America*. Part II. American Mathematical Society, Providence. 1989.

Eco, Umberto. *Foucault's Pendulum*. Translated by William Weaver. Ballantine Books, New York. 1989.

Eddington, Sir Arthur Stanley. *The Nature of the Physical World*. Cambridge University Press. 1928.

Eddington, Sir Arthur Stanley. *The Philosophy of Physical Science*. The University of Michigan Press, Ann Arbor. 1958.

Edmonds, Charles. *Poetry of the Anti-Jacobin*. Sampson Low, Marstow, Searle, & Rivington, London. 1890.

Edwards, Tryon. *The New Dictionary of Thoughts*. Standard Book Company. 1960.

Egler, Frank E. *The Way of Science*. Hafner Publishing Company, New York. 1970.

Egrafov, M. *Mathematics Magazine*. Volume 65, Number 5. December 1992.

Einstein, Albert. *Sidelights on Reality*. E.P. Dutton, New York. 1922.

Eldridge, Paul. *Maxims for a Modern Man*. Thomas Yoseloff: Publisher, New York. 1965.

Eliot, George. *Adam Bede*. Houghton Mifflin Company, Boston. 1907.

Eliot, George. *Felix Holt, The Radical*. Clarendon Press, Oxford. 1980.

Eliot, George. *The Works of George Eliot*. William Blackwood and Sons, Edinburgh. No date.

Ellis, Havelock. *The Dance of Life*. Houghton Mifflin Company, Boston. 1923.

Emerson, Ralph Waldo. *Society and Solitude*. Houghton, Mifflin and Company, Boston. 1898.

Emerson, Ralph Waldo. *The Complete Essays and other Writings of Ralph Waldo Emerson*. The Modern Library, New York. 1950.

Emerson, Ralph Waldo. *The Portable Emerson*. The Viking Press, New York. 1946.

Emerson, Ralph Waldo. *The Selected Writings of Ralph Waldo Emerson*. Modern Library, New York. 1950.

Emerson, Ralph Waldo. *The Works of Ralph Waldo Emerson*. Volume VI. Houghton Mifflin Company, Boston. 1909.

Esar, Evan. *20,000 Quips & Quotes*. Barnes & Noble Books, New York. 1968.

Esar, Evan. *Esar's Comic Dictinary*. Doubleday & Company Inc., Garden City. 1983.

Euclid. *The Thirteen Books of Euclid's Elements* in *Great Books of the Western World*. Translated by Sir Thomas L. Heath. Volume 11. Encyclopædia Britannica, Inc., Chicago. 1952.

Euripides. *The Plays of Euripides* in *Great Books of the Western World*.

Translated by Edward P. Coleridge. Volume 5. Encyclopædia Britannica, Inc., Chicago. 1952.

Evans, Bergen. *The Natural History of Nonsense*. A.A. Knopf, New York. 1946.

Eves, Howard W. *In Mathematical Circles*. Volume 2. Prindle, Weber & Schmidt, Inc., Boston. 1969.

Eves, Howard W. *Mathematical Circles Squared*. Prindle, Weber & Schmidt, Inc., Boston. 1972.

Fabing, Harold and Marr, Ray. *Fischerisms*. C.C. Thomas, Springfield. 1937.

Fabre, J. Henri. *The Life of the Spider*. Translated by Alexander Teixeira de Mattos. Doss, Mead and Company, New York. 1921.

Fadiman, Clifton. *The Mathematical Magpie*. Simon and Schuster, New York. 1962.

Fairbairn, A.M. *The Philosophy of the Christian Religion*. The Macmillan Company, New York. 1909.

Felix, Lucienne. *The Modern Aspects of Mathematics*. Translated by Julius H. Hlavaty and Fancille H. Hlavaty. Basic Books, New York. 1960.

Feller, William. *An Introduction to Probability Theory and Its Applications*. Volume 1. John Wiley & Sons, Inc., New York. 1960.

Feynman, Richard. *The Character of Physical Law*. British Broadcasting Corporation, London. 1965.

Feynman, Richard P., Leighton, Robert B. and Sands, Matthew. *The Feynman Lectures on Physics*. Volume I. Addison-Wesley Publishing Company, Reading. 1964.

Feynman, Richard P. *The Feynman Lectures on Physics*. Volume III. Addison-Wesley Publishing Company, Reading. 1964.

FitzGerald, Edward. *The Rubáiyát of Omar Khayyám*. Walter J. Black, Inc., New York. 1932.

Flammarion, Camille. *Popular Astronomy*. Chatto & Windus, Piccadilly. 1894.

Fogelfroe, Professor. 'Professor Fogelfroe' in *Mathematical Magazine*. Volume 69, Number 2. April 1996.

Fort, Tomlinson. 'Mathematics and the Sciences' in *The American Mathematical Monthly*. Volume 47, Number 9. November 1940.

Fosdick, Harry Emerson. *On Being a Real Person*. Harper & Brothers, Publishers, New York. 1943.

Fourier, Jean Baptiste Joseph. *Analytical Theory of Heat*. Translated, with notes, by Alexander Freeman. The University Press, Cambridge. 1878.

Francis, Richard L. 'On Coloring a Map' in *Mathematics Magazine*. Volume 65, Number 5. December 1990.

Frege, G. *The Foundations of Arithmetic*. Northwestern University Press, Evanston. 1968.

French, A.P. *Einstein: A Centenary Volume*. Harvard University Press, Cambridge. 1979.

Freud, Sigmund. *On Narcissism* in *Great Books of the Western World*. Translated by Cecil M. Baines. Volume 54. Encyclopædia Britannica, Inc., Chicago. 1952.

Frost, Robert. *The Poetry of Robert Frost*. Holt, Rinehart, and Winston, New York. 1965.

Froude, James Anthony. *Short Studies on Great Subjects*. E.P. Dutton, New York. 1930–31.

Fuchs, Walter R. *Mathematics for the Modern Mind*. Translated by Dr. H.A. Holstein. The Macmillan Company, New York. 1967.

Funk, Wilfred. 'Higher Mathematics' in *The Mathematics Teacher*. Volume XXIX, Number 1. January 1936.

Futrelle, Jacques. *Best "Thinking Machine" Detective Stories*. Dover Publications, Inc., New York. 1973.

Galilei, Galileo. 'Dialogues Concerning the Two New Sciences' in *Great Books of the Western World*. Translated by Henry Crew and Alfonso de Salvio. Volume 28. Encyclopædia Britannica, Inc., Chicago. 1952.

Gamow, George. *Biography of Physics*. Harper & Row Publishers, New York. 1961.

Gamow, George. *My World Lines*. The Viking Press, New York. 1970.

Gardner, Martin. *Mathematical Circus*. Alfred A. Knopf, New York. 1979.

Gass, Fredrick. 'Constructive Ordinal Notation Systems' in *Mathematics Magazine*. Volume 57, Number 3. May 1984.

Gates, Bill. *U.S. News and World Report*. 15 February, 1993.

Gibbon, Edward. *The Decline and Fall of the Roman Empire* in *Great Books of the Western World*. Volume 41. Encyclopædia Britannica, Inc., Chicago. 1952.

Gilbert, William and Sullivan, Arthur. *The Mikado, and Other Plays*. Modern Library, New York. 1917.

Ginsey, Gurney. 'Numbers' in *Mathematics Magazine*. Volume 38, Number 3. May 1965.

Glanvill, Joseph. *The Vanity of Dogmatizing*. Columbia University Press, New York. 1931.

Gleason, Andrew. 'Evolution of an Active Mathematical Theory' in *Science*. Volume 145. 31 July 1964.

Gleick, James. *Chaos: Making a New Science*. Viking Press, New York. 1987.

Goethe, Johann Wolfgang von. *Goethe's Poems and Aphorisms*. Oxford University Press, New York. 1932.

Goethe, Johann Wolfgang von. *Criticisms, Reflections, and Maxims of Goethe*. Translated by W.B. Ronnfeldt. The Walter Scott Publishing Co., Ltd., London. No date.

Goethe, Johann Wolfgang von. *Faust* in *Great Books of the Western World*. Translated by George Madison Priest. Volume 47. Encyclopædia Britannica, Inc., Chicago. 1952.

Good, Irving John. *The Scientist Speculates*. Basic Books, Inc., New York. 1962.

Graham, L.A. *Ingenious Mathematical Problems and Methods*. Dover Publications, Inc., New York. 1959.

Graham, Ronald L., Knuth, Donald E. and Patashnik, Oren. *Concrete Mathematics*. Addison-Wesley, Reading. 1989.

Graves, Robert Perceval. *Life of Sir William Rowan Hamilton*. Volume III. Hodges, Figges & Co., Dublin. 1882–1889.

Gray, Jeremy. 'Did Poincaré say "Set theory is a Disease?"' in *The Mathematical Intelligencer*. Volume 12, Number 1. 1991.

Green, Celia. *The Decline and Fall of Science*. Hamilton, London. 1976.

Greer, Scott. *The Logic of Social Inquiry*. Aldine Publishing Co., Chicago. 1973.

Growney, JoAnne S. 'Are Mathematics and Poetry Fundamentally Similar?' in *The American Mathematically Monthly*. Volume 99, Number 2. February 1992.

Gudder, Stanley. *A Mathematical Journey*. McGraw-Hill Book Company, New York. 1976.

Guillen, Michael. *Bridges to Infinity*. Jeremy P. Tarcher, Inc., Los Angeles. 1983.

Gullberg, Jan. *Mathematics from the Birth of Numbers*. W.W. Norton & Company, New York. 1997.

Gunther, John. *Taken at the Flood: The Story of Albert D. Lasker*. Harper & Brothers Publishers, New York. 1960.

Guy, Richard K. *Unsolved Problems in Number Theory*. Second Edition. Springer-Verlag, New York. 1994.

Hales, Stephen. *Vegetable Staticks*. The Scientific Book Guild, London. 1961.

Hall, G. Stanley. *Educational Problems*. Volume II. D. Appleton and Company, New York. 1911.

Halmos, Paul. *A Hilbert Space Problem Book*. D. Van Nostrand Company, Inc., Princeton. 1967.

Halmos, Paul. 'Mathematics as a Creative Art' in *American Scientist*. Volume 56, Winter, 1968.

Halmos, Paul. 'How to Write Mathematics' in *L'Enseignement Mathématique*. Volume 16. 1970.

Halsted, George Bruce. 'Biology and Mathematics' in *Science*. Volume XXII, Number 554. Friday, August 11, 1905.

Hardy, G.H. *A Mathematician's Apology*. Cambridge University Press, Cambridge. 1967.

Hardy, G.H. 'An Introduction to the Theory of Numbers' in *Bulletin of the American Mathematical Society*. Volume 35. 1929.

Hardy, Thomas. *Jude the Obscure*. With introduction by Terry Eagleton. Macmillan London Ltd., London. 1974.

Harris, W.T. *Psychologic Foundations of Education*. D. Appleton and Company, New York. 1898.

Harrison, Edward. *Masks of the Universe*. Macmillan Publishing Company, New York. 1985.

Hauffman, Paul. 'The Man Who Loves Only Numbers' in *The Atlantic Magazine*. Volume 260, Number 5. November 1987.

Hawking, Stephen. *A Brief History of Time*. Bantam Books, Toronto. 1988.

Heaviside, Oliver. *Electromagnetic Theory*. Dover Publications, Inc., New York. 1950.

Heims, Steve J. *From Mathematics to the Technologies of Life and Death*. The MIT Press, Cambridge. 1984.

Heinlein, Robert A. *Expanded Universe*. Grosset & Dunlap Publishers, New York. 1980.

Heinlein, Robert A. *The Cat Who Walks Through Walls*. Putman, New York. 1985.

Heinlein, Robert A. *The Number of the Beast*. New English Library, London. 1980.

Heinlein, Robert A. *Time Enough for Love*. G.P. Putnam's Sons, New York. 1973.

Heisenberg, Werner. *The Physical Principles of the Quantum Theory*. Translated by Carl Ekhart and Frank C. Hoyt. The University of Chicago Press. 1930.

Heller, Joseph. *Catch-22*. Dell Publishing Co., Inc., New York. 1955.

Helmholtz, H. *Popular Lectures on Scientific Subjects*. A. Appleton and Company, New York. 1885.

Henry, O. *Tales of O. Henry*. Doubleday, Garden City. 1969.

Herbert, George. *The Works of George Herbert*. Oxford at the Clarendon Press, London. 1941.

Herodotus. *The History of Herodotus* in *Great Books of the Western World*. Translated by George Rawlinson. Volume 6. Encyclopædia Britannica, Inc., Chicago. 1952.

Herschel, Sir John. *Outlines of Astronomy*. P.F. Collier & Son, New York. 1902.

Hertz, H. *Electric Waves; Being Researches on the Propagation of Electric Action with Finite Velocity through Space*. Macmillan & Co., London. 1893.

Heywood, Robert B. *The Works of the Mind*. The University of Chicago Press, Chicago. 1947.

Hilbert, David. 'Hilbert: Mathematical Problems' in *Bulletin of the American Mathematical Society*. Volume 8, 2nd Series. July 1902.

Hill, Thomas. 'The Imagination in Mathematics' in *North American Review*. Volume 85. 1857.

Hobbes, Thomas. *Leviathan*. E.P. Dutton and Company, Inc., New York. 1950.

Hodges, Wilfred. *Building Models by Games*. Cambridge University Press, Cambridge. 1985.

Hodnett, Edward. *The Art of Problem Solving*. Harper and Row, Publishers, New York. 1955.

Hofstadter, Douglas. *Gödel, Escher, Bach: An Eternal Golden Braid*. Basic Books, New York. 1979.

Hogben, Lancelot. *Mathematics for the Million*. W.W. Norton & Company, Inc., New York. 1964.

Holmes, Oliver Wendell. *Pages from an Old Volume of Life*. The Riverside Press, Cambridge. 1895.

Holmes, Oliver Wendell. *The Autocrat of the Breakfast-Table*. Houghton Mifflin, Boston. 1894.

Holmes, Oliver Wendell. *The Poet at the Breakfast-Table*. James R. Osgood and Company, Boston. 1872.

Holmes, Oliver Wendell, Jr. *The Common Law*. Little, Brown and Company, Boston. 1923.

Holt, Michael. *Mathematics in Art*. Van Nostrand Reinhold Company, New York. 1971.

Honsberger, Ross. *Mathematical Morsals*. The Mathematical Association of America. 1978.

Hopkins, Harry. *The Numbers Game: The Bland Totalitarianism*. Martin Secker & Warburg Ltd., London. 1973.

Housman, Alfred Edward. *The Collected Poems of A.E. Housman*. At the Alden Press, Oxford. 1939.

Howson, A.G. *Developments in Mathematical Education: Proceedings of the Second International Congress on Mathematical Education*. At the University Press, Cambridge. 1973.

Hoyle, Fred. *Of Men and Galaxies*. University of Washington Press, Seattle. 1964.

Hubbard, Elbert. *The Roycroft Dictionary*. The Roycrofters, East Aurora. 1914.

Hugo, Victor. *The Toilers of the Sea*. Volumes III–IV. Translated by Mary W. Artois. The Rittenhouse Press, Philadelphia. 1892.

Huntington, Edward V. *The Continuum*. Harvard University Press. 1917.

Huntly, Henry Edwards. *The Divine Proportion: a Study in Mathematical Beauty*. Dover Publications, New York. 1970.

Huxley, Aldous. *After Many a Summer Dies the Swan*. Harper & Brothers Publishers, New York. 1939.

Huxley, Aldous. *Brave New World*. Harper & Row, Publishers, New York. 1946.

Huxley, Aldous. *Jesting Pilate*. Chatto & Windus, London. 1926.

Huxley, Aldous. *Music at Night and Other Essays*. The Fountain Press, New York. 1931.

Huxley, Aldous. *Stories, Essays, and Poems*. J.M. Dent, London. 1939.

Huxley, Thomas. *Collected Essays*. Volume VIII. Greenwood Press Publications, New York. 1968.

Huxley, Thomas. *Lay Sermons, Addresses, and Reviews*. D. Appleton & Company, New York. 1871.

Huxley, Thomas. *Man's Place in Nature*. University of Michigan Press, Ann Arbor. 1959.

Huxley, Thomas H. *Science and Education*. D. Appleton and Company, New York. No date.

Huxley, Thomas. 'Scientific Education: Notes of an After Dinner Speech' in *Macmillan's Magazine*. Volume XX. July 1869.

Iannelli, Richard. *The Devil's New Dictionary*. Citadel Press, Secaucus. 1983.

Infeld, Leopold. *Quest—An Autobiography*. Chelsea Publishing Company, New York. 1980.

Inge, William Ralph. *A Rustic Moralist*. G.P. Putnam's Sons, New York. 1937?.

Jeans, James. *The Mysterious Universe*. The Macmillan Company, New York. 1932.

Jeffers, Robinson. *The Beginning and the End*. Random House, New York. 1954.

Jeffreys, Sir Harold and Swirles, Bertha. *Methods of Mathematical Physics*. At the University Press, Cambridge. 1956.

Jerome, Jerome K. *The Idle Thoughts of an Idle Fellow*. H. Altemus, Philadelphia. 1890.

Jevons, W. Stanley. *The Principles of Science*. Macmillan and Co., London. 1887.

Johnson, Samuel. *Johnsonian Miscellanies*. Volume I. Arranged and edited by George Birkbeck Hill. Harper and Brothers, New York. 1897.

Johnston, Alva. *The Legendary Mizners*. Farrar, Strauss & Young, New York. 1953.

Jones, Raymond F. *The Non-Statistical Man*. Belmont Productions, Inc., New York. 1964.

Jonson, Ben. *Volpone*. Yale University Press, New Haven. 1962.

Joseph, Margaret. 'The Future of Geometry' in *The Mathematics Teacher*. Volume XXIX, Number 1. January 1936.

Juster, Norton. *The Phantom Tollbooth*. Epstein & Carroll Associates, Inc., New York. 1962.

Kac, Mark. *Enigmas of Chance*. Harper & Row, New York. 1985.

Kamath, B.A. 'The end of Π' in *The Physics Teacher*. May 1979.

Kant, Immanuel. *The Critique of Judgment* in *Great Books of the Western World*. Translated by James Creed Meredoth. Volume 42. Encyclopædia Britannica, Inc., Chicago. 1952.

Kaplan, Abraham. *The Conduct of Inquiry*. Chandler Publishing Co., San Francisco. 1964.

Karpinski, Louis C. *Bibliography of Mathematical Works Printed in America Through 1850*. The University of Michigan Press, Ann Arbor. 1940.

Kasner, Edward and Newman, James. *Mathematics and the Imagination*. Simon and Schuster, New York. 1967.

Keller, Helen. *The Story of My Life*. Grosset & Dunlap Publications, New York. 1902.

Kelly-Bootle, Stan. *The Devil's DP Dictionary*. McGraw-Hill Book Co., Inc., New York. 1981.

Keyser, Cassius J. *Mathematics and the Question of Cosmic Mind with Other Essays*. Scripta Mathematica, New York. 1935.

Keyser, Cassius J. *Mathematical Philosophy*. E.P. Dutton & Company, New York. 1922.

Keyser, Cassius J. *Mole Philosophy & Other Essays*. E.P. Dutton & Company, New York. 1927.

Keyser, Cassius J. *The Pastures of Wonder*. University Press, New York. 1929.

Keyser, Cassius J. 'The Universe and Beyond' in *Hibbert Journal*. Volume 2. 1903–1904.

King, Jerry P. *The Art of Mathematics*. Plenum Press, New York. 1992.

Kipling, Rudyard. *Rudyard Kipling's Verse*. Doubleday and Co., Inc., New York. 1940.

Kipling, Rudyard. *Rudyard Kipling's Verse: Inclusive Edition*. Hodder and Stoughton Ltd., London. 1912.

Kirkpatrick, Larry D. 'Numerical Sprites Indeed' in *The Physics Teacher*. November 1978.

Klein, Felix. 'Arithmetizing of Mathematics' in *Bulletin of the American Mathematical Society*. Volume 11, Number 8, 2nd series. May 1896.

Klein, Felix. *Elementary Mathematics from an Advanced Standpoint*. Dover Publications, New York. No date.

Kleiner, Israel. 'Thinking the Unthinkable: The Story of Complex Numbers' in *Mathematics Teacher*. Volume 81, Number 7. October 1988.

Kline: *Mathematics: A Cultural Approach*. Addison-Wesley Publishing Company, Inc., Reading. 1962.

Kline, Morris. *Mathematics and the Physical World*. Doubleday & Co., Garden City. 1959.

Kline, Morris. *Mathematics in Western Culture*. Oxford University Press, London. 1953.

Kline, Morris. *Mathematics: The Loss of Certainty*. Oxford University Press, New York. 1980.

Kline, Morris. *Mathematical Thought from Ancient to Modern Times*. Oxford University Press, New York. 1972.

Kline, Morris. *Why the Professor Can't Teach*. St. Martin's Press, Inc., New York. 1978.

Kline, Morris. 'Mathematics: From Precision to Doubt in 100 Years' in *U.S. News and World Report*. January 26, 1981.

Kline, Morris. 'Projective Geometry' in *Scientific American*. Volume 192, Number 1. January 1955.

Koestler, A. *The Act of Creation*. Pan Books, London. 1964.

Kolmogorov, A.N. *Foundations of the Theory of Probability*. Chelsea Publishing Co., New York. 1956.

Kürschak, Jozef. *Hungarian Problem Book*. Volume 1. Mathematical Association of America, Washington, D.C. 1963.

Kyburg, H.E., Jr. and Smokler, H.E. (Editors). *Studies in Subjective Probability*. Wiley, New York. 1964.

La Touche, Mrs. 'Gleanings Far and Near' in *Mathematical Gazette*. Volume 12, Number 170. May 1924.

Lakatos, Imre. *Mathematics, Science and Epistemology*. Volume II. Cambridge University Press, Cambridge. 1978.

Langer, R.E. 'The Things I Should Have Done, I Did Not Do' in *The American Mathematical Monthly*. Volume 59. September 1952.

Langer, Susan. *Philosophy in a New Key*. Harvard University Press, Cambridge. 1963.

Laplace, Pierre Simon. *Oeuvres Complètes de Laplace*. Gauthier-Villars, Paris. 1886.

Laplace, Pierre Simon. *Philosophical Essays on Probabilities*. Translated from the fifth French edition of 1825 by Andrew I. Dale. Springer-Verlag New York, Inc., New York. 1995.

Leacock, Stephen. *Literary Lapses*. John Lane Co., New York. 1914.

Lebowitz, Fran. *Social Studies*. Pocket Books, New York. 1981.

Lec, Stanislaw J. *More Unkempt Thoughts*. Funk & Wagnalls, New York. 1968.

Legman, G. *The New Limerick*. Crown Publishers, New York. 1977.

Lehmer, Derrick Henry. 'Mechanized Mathematics' in *Bulletin of the American Mathematical Society*. Volume 72, Number 5. September 1966.

Lehmer, D.N. 'Hunting Big Game in the Theory of Numbers' in *Scripta Mathematica*. Volume 1. 1932.

Lehrer, Tom. 'The Derivative Song' in *The American Mathematical Monthly*. Volume 81, Number 5, May 1974.

Lehrer, Tom. 'There's a Delta For Every Epsilon' in *The American Mathematical Monthly*. Volume 81, Number 6. June–July 1974.

Leibniz, Gottfried Wilhelm. *Philosophical Papers and Letters*. The University of Chicago Press, Chicago. 1956.

Lem, Stanislaw. *The Cyberiad*. The Seaburg Press, New York. 1974.

Leslie, John. *Elements of Geometry*. Edinburgh. 1809.

Lewis, C.S. *The Pilgrim's Regress: An Allegorical Apology for Christianity, Reason and Romanticism*. Beerdmans, Grand Rapids. 1943.

Lewis, Gilbert N. *The Anatomy of Science*. Yale University Press, New Haven. 1926.

Lewis, Sinclair. *Arrowsmith*. The Modern Library, New York. 1925.

Lichtenberg, George C. *Lichtenberg: Aphorisms & Letters*. Translated by Franz Mautner and Henry Hatfield. Jonathan Cape, London. 1969.

Lieber, Lillian R. *The Education of T.C. MITS*. W.W. Norton & Co., Inc., New York. 1944.

Lillich, Robert. 'My Fair Physicist' in *The Physics Teacher*. December 1968.

Lindsay, R.B. 'On the Relation of Mathematics and Physics' in *Scientific Monthly*. Volume 59, Number 6. December 1944.

Lindsay, Vachel. *The Congo and Other Poems*. Macmillan, New York. 1914.

Lippmann, Walter. *Liberty and the News*. Transaction Publishers, New Brunswick. 1995.

Lipschitz, R. 'Letter to the Editor' in *Annals of Mathematics*. 2nd Series. Volume 69. 1959.

Locke, John. *An Essay Concerning Human Understanding* in *Great Books of the Western World*. Volume 35. Encyclopædia Britannica, Inc., Chicago. 1952.

Locke, William J. *The House of Baltazar*. John Lane Company, New York. 1920.

Longair, Malcolm. 'Quasi-Stellar Radio Sources' in *Contemporary Physics*. Volume 8, Number 4. July 1967.

Longfellow, Henry Wadsworth. *The Poems of Longfellow*. Random House, Inc., New York. 1944.

Lover, Samuel. *Rory O'More*.

Lucretius. *On the Nature of Things* in *Great Books of the Western World*. Translated by H.A.J. Munro. Volume 12. Encyclopædia Britannica, Inc., Chicago. 1952.

Luminet, Jean-Pierre. *Black Holes*. University Press, Cambridge. 1987.

Mac Lane, Saunders. 'Of Course and Courses' in *The American Mathematical Monthly*. Volume 61. March 1954.

MacGowan, Roger A. and Ordway, Frederick I., III. *Intelligence in the Universe*. Prentice Hall, Englewood Cliffs. 1966.

Mach, Ernst. *The Science of Mechanics*. The Open Court Publishing Company, La Salle. 1942.

Machover, Maurice. 'Ode to the Continuum Hypothesis' in *Mathematics Magazine*. Volume 50, Number 2. March 1977.

Mackay, Charles. *The Poetical Works of Charles Mackay*. G. Routledge, London. 1857.

Manin, Yu. I. *A Course in Mathematical Logic*. Springer-Verlag, New York. 1977.

Mann, Lee. 'The Digit it is!' in *Arithmetic Teacher*. Volume 13, Number 8. December 1966.

Mann, Thomas. *The Magic Mountain*. Translated by H.T. Lowe-Porter. Alfred A. Knopf, New York. 1946.

Manning, Henry P. *The Fourth Dimension Simply Explained*. Dover Publications, Inc., New York. 1960.

Maor, Eli. *To Infinity and Beyond: A Cultural History of the Infinite*. Birkhäuser, Boston. 1987.

Marchand, Leslie A. *Byron's Letters and Journals*. Volume 3. Harvard University Press, Cambridge. 1974.

Marsden, Jerrold E. and Tromba, Anthony J. *Vector Calculus*. W.H. Freeman and Company, San Francisco. 1981.

Marvell, Andrew. *The Poetical Works of Andrew Marvell*. Alexander Murray, London. 1870.

Mason, Alpheus T. *Brandeis: A Free Man's Life*. The Viking Press, New York. 1946.

Mathematical Sciences Education Board. *Everybody Counts: A Report to the Nation on the Future of Mathematics Education*. National Academy Press, Washington, D.C. 1989.

Mathematical Sciences Education Board. *Reshaping School Mathematics*. National Academy Press, Washington, D.C. 1990.

Mautner, Franz H. and Hatfield, Henry. *The Lichtenberg Reader*. Beacon Press, Boston. 1959.

Maxwell, James Clerk. *The Life of James Clerk Maxwell*. Macmillan and Co., London. 1882.

Maxwell, James Clerk. 'Elements of Natural Philosophy' in *Nature*. Volume 7. March 27, 1873.

McCormack, Thomas J. 'On the Nature of Scientific Law and Scientific Explanation' in *The Monist*. Volume 10. 1899–1900.

Mehra, Jagdish. *The Physicist's Conception of Nature*. Dordrecht, Boston. 1973.

Mellor, J.W. *Higher Mathematics for Students of Chemistry and Physics*. Dover Publications, New York. 1955.

Melrose, A.R. *The Pooh Dictionary*. Dutton Books, New York. 1995.

Melville, Herman. *Moby Dick* in *Great Books of the Western World*. Volume 48. Encyclopædia Britannica, Inc., Chicago. 1952.

Mencke, J.B. *The Charlatanry of the Learned*. Translated by Francis E. Litz. Alfred A. Knopf, New York. 1937.

Merriman, Gaylord M. *To Discover Mathematics*. John Wiley & Sons, Inc., New York. 1942.

Merz, J.T. *A History of European Thought in the Nineteenth Century*. Volume 1. Dover Publications, New York. 1965.

Meyer, Walter. 'Missing Dimensions of Mathematics' in *Humanistic Mathematics Network Journal*. Number 11. February 1995.

Michelmore, Peter. *Einstein, Profile of the Man*. Dodd, Mead Publishers, New York. 1962.

Mikes, George. *How to Be an Alien*. Basic Books, Inc., New York. 1964.

Mill, John Stuart. *On Liberty*. The Bobbs-Merrill Company, Inc., Indianapolis. 1956.

Mill, John Stuart. *System of Logic*. Longmans, Green, Reader & Dyer, London. 1868.

Milne, A.A. *The House at Pooh Corner*. E.P. Dutton & Co., Inc., New York. 1928.

Minkowski, Hermann. *Diophantische Approximationen: eine Einführung in die Zahlentheorie, von Hermann Minkowski*. B.G. Teubner, Leipzig. 1907.

Montague, William Pepperell. 'The Einstein Theory and a Possible Alternative' in *Philosophical Review*. Volume 33, Number 194. March 1924.

Mordell, Louis Joel. *Reflections of a Mathematician*. Canadian Mathematical Congress, Montreal. 1959.

More, Louis Trenchard. *The Limitations of Science*. Henry Holt and Company, New York. 1915.

Morgan, Robert. 'Π' in *Poetry*. Volume CLXI, Number 4. January 1993.

Morley, Christopher. *The Ballad of New York, New York and Other Poems 1930–1950*. Doubleday, New York. 1950.

Morley, Christopher. *Translations from the Chinese*. Doubleday Page and Company, Garden City. 1927.

Morton, H.V. *In Search of Scotland*. Methuen & Co., Ltd., London. 1949.

Moultrie, John. *Poems*. Macmillan, London. 1876.

Moultrie, John. *The Dream of Life*. William Pickering, London. 1843.

Nash, Ogden. *Parents Keep Out*. Little, Brown and Company, Boston. 1951.

Newman, James R. *The World of Mathematics*. Volumes I, II, III, IV. Simon and Schuster, New York. 1956.

Newman, M.H.A. 'What is Mathematics?' in *Mathematical Gazette*. Volume 43, Number 345. October 1959.

Newton, Sir Isaac. *Mathematical Principles of Natural Philosophy* in *Great Books of the Western World*. Volume 34. Encyclopædia Britannica, Inc., Chicago. 1952.

Newton, Sir Isaac. *Opticks*. Dover Publications, Inc., New York. 1952.

Nicomachus of Gerasa. *Introduction to Arithmetic*. Translated by Martin Luther D'Ooge. The Macmillan Company, New York. 1926.

Nietzsche, Friedrich. *Human, All-Too-Human*. University of Nebraska Press, Lincoln. 1984.

O'Brien, Katharine. '∀ and ∃' in *The Mathematical Magazine*. Volume 55, Number 1. January 1982.

O'Brien, Katharine. 'Einstein and the Ice-Cream Cone' in *The Mathematics Teacher*. April 1968.

Olson, Richard. *Scottish Philosophy and British Physics: 1750–1880*. Princeton University Press, Princeton. 1975.

Oman, John. *The Natural & the Supernatural*. The Macmillan Company, New York. 1931.

Oppenheimer, Julius Robert. 'The Scientific Foundations for World Order' in *Foundations for World Order*. The University of Denver Press, Denver. 1949.

Oppenheimer, Julius Robert. 'The Tree of Knowledge' in *Harper's Magazine*. Volume 217, Number 1301. October 1955.

Ore, Oystein. *Niels Henrik Abel: Mathematician Extraordinary*. University of Minnesota Press, Minneapolis. 1957.

Osserman, Robert. *Poetry of the Universe*. Anchor Books, New York. 1995.

Page, Ray. *Quote*. Volume 53, Number 20. May 14, 1967.

Pagels, Heinz R. *The Dreams of Reason*. Simon and Schuster, New York. 1988.

Pagels, Heinz. *The Cosmic Code*. Simon and Schuster, New York. 1982.

Paget, R.L. *The Poetry of American Wit and Humor*. L.C. Page and Company, Boston. 1899.

Papert, Seymour. *Mindstorms*. Basic Books, Inc., New York. 1980.

Parker, Francis W. *Talks on Pedagogics*. E.L. Kellogg & Co., New York. 1894.

Parrott, E.O. *The Penguin Book of Limericks*. A. Lane, London. 1983.

Pascal, Blaise. *Pensées* in *Great Books of the Western World*. Translated by W.F. Trotter. Volume 33. Encyclopædia Britannica, Inc., Chicago. 1952.

Paulos, John Allen. *Innumeracy*. Hill and Wang, New York. 1988.

Pearson, Karl. *The Grammar of Science*. J.M. Dent and Sons Ltd., London. 1937.

Pétard, H. 'A Brief Dictionary of Phrases Used in Mathematical Writing' in *The American Mathematical Monthly*. Volume 73. 1966.

Peirce, Benjamin. 'Linear Associative Algebra' in *American Journal of Mathematics*. Volume 4. 1881.

Peirce, Charles Sanders. 'The Architecture of Theories' in *The Monist*. Volume 1, Number 2. January 1891.

Peirce, Charles Sanders. *Chance, Love, and Logic*. Harcourt, Brace & Company, Inc., New York. 1923.

Peirce, Charles Sanders. *Writings of Charles Sanders Peirce*. Volume 3, 1872–1878. Indiana University Press, Bloomington. 1958.

Peirce, Charles Sanders. 'The Regenerated Logic' in *The Monist*. Volume 7, Number 1. 1896.

Pétard, H. 'A Brief Dictionary of Phrases Used in Mathematical Writing' in *The American Mathematical Monthly*. Volume 73, Number 2. February 1966.

Pétard, H. 'A Contribution to the Mathematical Theory of Big Game Hunting' in *The American Mathematical Monthly*. Volume 45. 1938.

Peterson, Ivars. *Islands of Truth: A Mathematical Mystery Cruise.* W.H. Freeman and Company, New York. 1990.

Peterson, Ivars. *The Mathematical Tourist.* W.H. Freeman and Company, New York. 1988.

Picard, Émile. *Traite d'Analyse.* Gauthier-Villars et fils, Paris. 1893–1901.

Pickover, Clifford A. *Keys to Infinity.* Wiley & Sons, New York. 1995.

Pierpont, James. 'The History of Mathematics in the Nineteenth Century' in *Bulletin of the American Mathematical Society.* 2nd Series. Volume 11. 1904–1905.

Pierpont, James. 'Mathematical Rigor' in *Bulletin of the American Mathematical Society.* Volume 34. January–February 1928.

Pirsig, Robert M. *Zen and the Art of Motorcycle Maintenance.* William Morrow & Co., Inc., New York. 1974.

Plato. *Charmides* in *Great Books of the Western World.* Translated by Benjamin Jowett. Volume 7. Encyclopædia Britannica, Inc., Chicago. 1952.

Plato. *Gorgias* in *Great Books of the Western World.* Translated by Benjamin Jowett. Volume 7. Encyclopædia Britannica, Inc., Chicago. 1952.

Plato. *Philebus* in *Great Books of the Western World.* Translated by Benjamin Jowett. Volume 7. Encyclopædia Britannica, Inc., Chicago. 1952.

Plato. *Statesman* in *Great Books of the Western World.* Translated by Benjamin Jowett. Volume 7. Encyclopædia Britannica, Inc., Chicago. 1952.

Plato. *The Republic* in *Great Books of the Western World.* Translated by Benjamin Jowett. Volume 7. Encyclopædia Britannica, Inc., Chicago. 1952.

Pliny the Elder. *Natural History.* The Harvard University Press, Cambridge. 1962.

Plotinus. *The Six Enneads* in *Great Books of the Western World.* Translated by Stephen MacKenna and B.S. Page. Volume 17. Encyclopædia Britannica, Inc., Chicago. 1952.

Plotz, Helen. *Imagination's Other Place.* Thomas Y. Crowell Company, New York. 1955.

Poe, Edgar Allan. *The Complete Edgar Allan Poe Tales.* Avenel Books, New York. 1981.

Poe, Edgar Allan. *Seven Tales.* Schocken Books, New York. 1971.

Pohl, Frederik. *The Coming of the Quantum Cats.* Bantam, New York. 1986.

Poincaré, Henri. *Science and Method.* Dover Publications, Inc., New York. No date.

Poincaré, Henri. *The Foundation of Science.* The Science Press, New York. 1921.

Poincaré, Henri. 'Sur les Équations aux Dérivées Partielles de la Physique Mathématique' in *American Journal of Mathematics.* Volume 12. 1890.

Poincaré, Henri. 'The Relation of Analysis and Mathematical Physics' in *Bulletin of the American Mathematical Society.* Volume IV. 1899.

Poincaré, Henri. 'Non-Euclidean Geometry' in *Nature*. Volume 46, 1891–1892.

Polanyi, Michael. *Personal Knowledge*. The University of Chicago Press, Chicago. 1958.

Pólya, G. *How to Solve It*. Princeton University Press, Princeton. 1948.

Ponomarev, L.I. *The Quantum Dice*. IOP Publishing, Bristol. 1993.

Pope, Alexander. *The Complete Poetical Works*. Houghton Mifflin and Company, New York. 1903.

Pottage, John. *Geometrical Investigations*. Addison-Wesley Publishing Company, Inc., Reading. 1983.

Price, Bartholomew. *A Treatise on Infinitesimal Calculus*. Volume III. Second Edition. At the Clarendon Press, Oxford. 1868.

Queneau, Raymond. *Pounding the Pavement, Beating the Bush, and Other Pataphysical Poems*. Translated by Teo Savory. Unicorn Press, Greensboro. 1985.

Quine, W.V.O. *Elementary Logic*. Harvard University Press, Cambridge. 1980.

Rankine, William J.M. *Song and Fables*. J. Maclehose, Glasgow. 1874.

Rebière, Alphonse. *Mathématiques et Mathématiciens: Pensées et Curiosités*. Nony, Paris. 1889.

Reichen, Charles-Albert. *A History of Astronomy*. Hawthorn Books Inc., New York. 1963.

Reichenbach, Hans. *The Philosophy of Space & Time*. Dover Publications, Inc., New York. 1958.

Reid, Constance. *Hilbert*. Springer-Verlag, New York. 1970.

Reid, Thomas. *Essays on the Intellectual Powers of Man*. Macmillan and Co., Limited, London. 1941.

Reynolds, Henry T. *Analysis of Nominal Data*. SAGE Publications, Beverly Hills. 1977.

Reznick, Bruce. 'A Set is A Set' in *Mathematics Magazine*. Volume 66, Number 2. April 1993.

Reznick, Bruce *et al.* 'Hooray for Calculus' in *Mathematical Magazine*. Volume 61, Number 3. June 1988.

Richardson, Moses. 'Mathematics and Intellectual Honesty' in *The American Mathematical Monthly*. Volume 59, Number 2. February 1952.

Ritchie, Robert W. *New Directions in Mathematics*. Prentice-Hall, Inc., Englewood Cliffs. 1963.

Robson, Ernest and Wimp, Jet. *Against Infinity*. Primary Press, Parker Ford. 1979.

Roe, E.D., Jr. 'A Generalized Definition of Limit' in *The Mathematics Teacher*. Volume III, Number 1. September 1910.

Rogers, Eric M. *Physics for the Inquiring Mind*. Princeton University Press, Princeton. 1960.

Romanoff, Alexis L. *Encyclopedia of Thoughts*. Ithaca Heritage Books, Ithaca. 1957.

Rosenbaum, R.A. 'Mathematics, the Artistic Science' in *Mathematical Teacher*. Volume 55, Number 7. November 1962.

Rosenblatt, Roger. *The Man in the Water*. Random House, New York. 1994.

Royce, Josiah. *The World and the Individual*. Macmillan & Co., London. 1901.

Rózsa, Péter. *Playing with Infinity*. Simon and Schuster, New York. 1962.

Rucker, Rudy. *Infinity and the Mind*. Birkhäuser, Boston. 1982.

Russell, Bertrand. *An Essay on the Foundations of Geometry*. At the University Press, Cambridge. 1897.

Russell, Bertrand. *Introduction to Mathematical Philosophy*. George Allen & Unwin, Ltd., London. 1920.

Russell, Bertrand. *Mysticism and Logic and Other Essays*. Longmans Green and Co., London. 1918.

Russell, Bertrand. *Portraits from Memory and Other Essays*. Simon and Schuster, New York. 1956.

Russell, Bertrand. *The ABC of Relativity*. Allen & Unwin, London. 1969.

Russell, Bertrand. *The Analysis of Matter*. Dover Publications, New York. 1954.

Russell, Bertrand. *The Autobiography of Bertrand Russell*. Little, Brown and Company, Boston. 1931.

Russell, Bertrand. *The Collected Stories of Bertrand Russell*. George Allen & Unwin Ltd., London. 1972.

Russell, Bertrand. *The Principles of Mathematics*. W.W. Norton & Company, Inc., New York. 1938.

Russell, Bertrand. 'Recent Work on the Principles of Mathematics' in *The International Monthly*. July–December 1901.

Sacks, Oliver. *The Man Who Mistook His Wife for a Hat and Other Clinical Tales*. Summit Books, New York. 1985.

Sagan, Carl. *Contact: A Novel*. Simon and Schuster, New York. 1985.

Sage, M. *Mrs. Piper and the Society for Psychical Research*. Scott-Thaw Co., New York. 1904.

Sandburg, Carl. *Harvest of Poems: 1910–1960*. Harcourt, Brace & World, Inc., New York. 1960.

Santayana, George. *Realm of Truth*. Charles Scribner's Sons, New York. 1938.

Santayana, George. *Some Turns of Thought in Modern Philosophy*. Charles Scribner's Sons, New York. 1933.

Sarton, George. *The Study of the History of Mathematics*. Dover Publications, Inc., New York. 1936.

Sawyer, W.W. *A First Look at Numerical Analysis*. Clarendon Press, Oxford. 1978.

Sawyer, W.W. *Prelude to Mathematics.* Penguin Books Ltd., Middlesex. 1960.

Sawyer, W.W. 'Algebra' in *Scientific American.* Volume 211, Number 3. September 1964.

Schaaf, William L. *Mathematics: Our Great Heritage.* Harper & Brothers, New York. 1948.

Schenck, Harold Jr. 'Wackyjabber' in *The Magazine of Fantasy and Science Fiction.* May 1960.

Schrödinger, Erwin. *Science and the Human Temperament.* W.W. Norton & Company, Inc., New York. 1935.

Schubert, Hermann. *Mathematical Essays and Recreations.* The Open Court Publishing Company, Chicago. 1903.

Serge, Corrado. 'On Some Tendencies in Geometric Investigations' in *Bulletin of the American Mathematical Society.* Volume 10. June 1904.

Serge, Lang. *The Beauty of Doing Mathematics.* Springer-Verlag, New York. 1985.

Shadwell, Thomas. *The Complete Works of Thomas Shadwell.* The Fortune Press, London. 1927.

Shakespeare, William. *Hamlet, Prince of Denmark* in *Great Books of the Western World.* Volume 27. Encyclopædia Britannica, Inc., Chicago. 1952.

Shakespeare, William. *The Merchant of Venice* in *Great Books of the Western World.* Volume 26. Encyclopædia Britannica, Inc., Chicago. 1952.

Shakespeare, William. *The Merry Wives of Windsor* in *Great Books of the Western World.* Volume 27. Encyclopædia Britannica, Inc., Chicago. 1952.

Shakespeare, William. *Othello, Moor of Venice* in *Great Books of the Western World.* Volume 27. Encyclopædia Britannica, Inc., Chicago. 1952.

Shakespeare, William. *The Taming of the Shrew* in *Great Books of the Western World.* Volume 27. Encyclopædia Britannica, Inc., Chicago. 1952.

Shaw, Bernard. *The Complete Plays of Bernard Shaw.* Odhams Press Limited, London.

Shaw, Bernard. *The Doctor's Dilemma.* Brentano's, New York. 1920.

Shaw, James Byrnie. *Lectures on the Philosophy of Mathematics.* The Open Court Publishing Company, Chicago. 1918.

Shelley, Percy Bysshe. *The Complete Poetical Works of Shelley.* Houghton Mifflin Co., Boston. 1901.

Shive, John N. and Weber, Robert L. *Similarities in Physics.* John Wiley & Sons, New York. 1982.

Sholander, Marlow. 'Envelopes and Nodes' in *Mathematics Magazine.* Volume 34, Number 2. November–December 1960.

Sholander, Marlow. 'Maybe' in *Mathematics Magazine.* Volume 35, Number 1. January 1962.

Siegel, Eli. *Damned Welcome.* Definition Press, New York. 1972.

Simmons, W.F.G. *The Mathematical Intelligencer*. Volume 13, Number 1. Winter 1991.

Smith, Adam. *The Money Game*. Random House, New York. 1968.

Smith, David Eugene. *The Poetry of Mathematics and Other Essays*. Scripta Mathematica, New York. 1934.

Smith, E.E. *Masters of the Vortex*. Pyramid, New York. 1968.

Smith, Henry J.S. 'Opening Address by the President Professor Henry J.S. Smith' in *Nature*. Section A, Volume 8. September 25, 1873.

Smith, Sydney. *The Letters of Sydney Smith*. Edited by Nowell C. Smith. At the Clarendon Press, Oxford. 1953.

Solzhenitsyn, Aleksandr. *The First Circle*. Harper and Row Publishers, New York. 1968.

Spencer-Brown, George. *Laws of Form*. Allen & Unwin, London. 1969.

Spengler, Oswald. *The Decline of the West*. Volume I. Alfred A. Knopf, New York. 1934.

Spiegel, M.R. 'Our Mathematical Alphabet' in *Mathematics Magazine*. Volume 57, Number 3. May 1984.

Stabler, E. Russell. 'An Interpretation and Comparison of Three Schools of Thought in the Foundation of Mathematics' in *The Mathematics Teacher*. Volume 26. January 1935.

Steen, Lynn Arthur. *Mathematics Today: Twelve Informal Essays*. Springer-Verlag, New York. 1978.

Steen, Lynn Arthur. *Mathematics Tomorrow*. Springer-Verlag, New York. 1981.

Steinbeck, John. *The Moon is Down*. Heinemann, London. 1942.

Stendhal. *The Life of Henri Brulard*. Translated by Catherine Alison Phillips. Vintage Books, New York. 1955.

Sterne, Lawrence. *Tristram Shandy*. The Modern Library, New York. 1928.

Stevenson, Robert Louis. *The Complete Poems of Robert Louis Stevenson*. C. Scribner's Sons, New York. 1923.

Stewart, Dugald. *The Collected Works of Dugald Stewart*. Volume IV. Edited by Sir William Hamilton, Bart. Thomas Constable and Co., Edinburgh. 1854.

Stewart, Ian. *Does God Play Dice?* Basil Blackwell Inc., Cambridge. 1990.

Stewart, Ian. *Nature's Numbers*. Basic Books, New York. 1995.

Stone, Marshall H. 'Mathematics and the Future of Science' in *Bulletin of the American Mathematical Association*. Volume 3, Number 2. March 1957.

Struik, Dirk J. *A Concise History of Mathematics*. Dover Publications, Inc., New York. 1948.

Sukoff, Albert. 'Lotsa Hamburgers' in *Saturday Review of the Society*. March 1973.

Sullivan, J.W.N. *Aspects of Science*. Jonathan Cape & Harison Smith, New York. 1923.

Sullivan, J.W.N. *The History of Mathematics in Europe*. Oxford University Press, London. 1924.

Swann, W.F.G. *The Architecture of the Universe*. The Macmillan Company, New York. 1934.

Swift, Jonathan. *Gulliver's Travels* in *Great Books of the Western World*. Volume 36. Encyclopædia Britannica, Inc., Chicago. 1952.

Swift, Jonathan. *Satires and Personal Writings*. Oxford University Press, New York. 1932.

Swift, Jonathan. *The Portable Swift*. Viking Press, New York. 1948.

Sylvester, James Joseph. *The Collected Mathematical Papers of James Joseph Sylvester*. Volumes II and III. At the University Press, Cambridge. 1908.

Sylvester, James Joseph. *Philosophical Magazine*. Volume 24. 1844.

Sylvester, James Joseph. *Philosophical Magazine*. Volume 31, 1866.

Sylvester, James Joseph. 'A Plea for the Mathematician' in *Nature*. Volume 1. December 30, 1869.

Synge, John L. 'The Life and Early Works of Sir William Rowan Hamilton' in *The Scripta Mathematica Studies Number 2*. Scripta Mathematica, New York. 1945.

Synge, J.L. 'Postcards on Applied Mathematics' in *The American Mathematical Monthly*. Volume 46, Number 3. March 1939.

Tarbell, Ida M. *The Ways of Woman*. The Macmillan Co., New York. 1916.

Taylor, E.G.R. *The Mathematical Practitioners of Tudor & Stuart England*. At the University Press, Cambridge. 1954.

Teller, Edward. *Conversations on the Dark Secrets of Physics*. Plenum Press, New York. 1991.

Teller, Edward. *The Pursuit of Simplicity*. Pepperdine University Press, Malibu. 1981.

Terence. *Phormio*. Translated by Frank O. Copley. The Liberal Arts Press, New York. 1958.

Thom, René. 'Modern Mathematics: An Educational and Professional Error?' in *American Scientist*. Volume 59. 1971.

Thompson, D'Arcy. *On Growth and Form*. Volume I. Cambridge University Press, London. 1959.

Thompson, Silvanus P. *Calculus Made Easy*. The Macmillan Company, New York. 1929.

Thompson, Silvanus P. *The Life of William Thomson Baron Kelvin of Largs*. Volume II. Macmillan and Company Ltd., London. 1910.

Thompson, W.R. *Science and Common Sense*. Longmans Green and Co., London. 1937.

Thoreau, Henry David. *A Week on the Concord and Merrimac Rivers*. Walter Scott Limited, London. 1889.

Thurber, James. *Many Moons*. Harcourt, Brace and Company, New York. 1943.

Titchener, Edward Bradford. *Systematic Psychology*. Cornell University Press, Ithaca. 1972.

Tolstoy, Leo. *Anna Karenina*. The Modern Library, New York. No date.

Tolstoy, Leo. *War and Peace* in *Great Books of the Western World*. Volume 51. Encyclopædia Britannica, Inc., Chicago. 1952.

Tomlinson, Fort. 'Mathematics and the Sciences' in *Scientific American*. Volume 47, Number 11. November 1940.

Tomlinson, Henry Major. *All Our Yesterdays*. Harper & Brothers Publishers, New York. 1930.

Trent, William Peterfield. *The Cambridge History of American Literature*. The Macmillan Company, New York. 1946.

Trevelyan, George Otto. *The Early History of Charles James Fox*. Harper & Brothers, Publisher, New York. 1900.

Trudeau, Richard J. *Dots and Lines*. The Kent State University Press. 1976.

Truesdell, C. *Essays in the History of Mechanics*. Springer-Verlag, New York. 1968.

Truesdell, Clifford. *Six Lectures on Modern Natural Philosophy*. Springer-Verlag, New York. 1966.

Tukey, John W. 'The Future of Data Analysis' in *The Annals of Mathematical Statistics*. Volume 33, Number 1. March 1962.

Tupper, Martin F. *Proverbial Philosophy: A Book of Thoughts and Arguments*. E.H. Butler & Co., Philadelphia. 1855.

Turner, H.H. *The Mathematical Gazette*. Volume VI, Number 100. October 1912.

Twain, Mark. *Mark Twain's Notebook*. Harper & Brothers Publishers, New York. 1935.

Twain, Mark. *Sketches Old and New*. American Publishing Company, Hartford. 1893.

Twain, Mark. *The Adventures of Huckleberry Finn*. Heritage Press, New York. 1940.

Tymoczko, Thomas. 'The Four Color Problems' in *Journal of Philosophy*. Volume 76. 1979.

Ulam, S.M. *Adventures of a Mathematician*. Charles Scribner's Sons, New York. 1976.

Ulam, Stanislaw. 'John von Neumann 1903–1957' in *Bulletin of the American Mathematical Society*. Volume 64, Number 3, part 2. May 1958.

University of Denver. *Foundations for World Order*. The University of Denver Press, Denver. 1949.

Updike, John. *Roger's Version*. Alfred A. Knopf, New York. 1986.

Updike, John. *Telephone Poles and Other Poems*. Alfred A. Knopf, New York. 1964.

Uspenskii, Petr Demianovich. *Tertium Organum*. Vintage Books, New York. 1970.

van der Waerden, B.L. *Science Awakening*. Oxford University Press, New York. 1961.

Van Dine, S.S. *The Bishop Murder Case*. C. Scribner's Sons, New York. 1929.

Veblen, Oswald. 'Geometry and Physics' in *Science*. Volume LVII, Number 1466. February 2, 1923.

Viereck, George Sylvester and Eldridge, Paul. *My First Two Thousand Years*. Sheridan House, Inc., New York. 1963.

Virgil. *The Eclogues*. Translated by C.S. Calverley. Heritage Press, New York. 1960.

Voltaire. *The Best Known Works of Voltaire*. Blue Ribbon Books, New York. 1927.

Voltaire. *The Portable Voltaire*. Viking Press, New York. 1949.

von Neumann, John. *Collected Works*. Volumes I and VI. Pergamon Press, New York. 1961.

Waismann, Friedrich. *Introduction to Mathematical Thinking*. Frederick Unger Publishing Company, New York. 1951.

Walker, Marshall. *The Nature of Scientific Thought*. Prentice-Hall, Inc., New Jersey. 1963.

Wall, H.S. *Creative Mathematics*. University of Texas Press, Austin. 1963.

Walter, Marion and O'Brien, Tom. 'Memories of George Pólya' in *Mathematics Teaching*. Volume 116. September 1986.

Walton, Izaak. *The Compleat Angler*. The Easton Press, Norwalk. 1948.

Warner, Sylvia Townsend. *Mr. Fortune's Maggot*. The Literary Guild, New York. 1927.

Weaver, Jefferson Hane. *The World of Physics*. Volumes II and III. Simon and Schuster, New York. 1987.

Weaver, Waren. 'Lewis Carroll: Mathematician' in *Scientific American*. Volume 194, Number 4. April 1956.

Webster, John. *The Duchess of Malfi*. Chatto and Windus, London. 1958.

Weil, André. *Œuvres Sciéntifiques*. Volume II. Springer-Verlag, New York. 1980.

Weil, André. 'Mathematical Teaching in Universities' in *The American Mathematical Monthly*. Volume 61, Number 1. January 1954.

Weil, André. 'The Future of Mathematics' in *The American Mathematical Monthly*. Volume 57, Number 5. May 1950.

Weinberg, Gerald M. *Rethinking System Analysis and Design*. Little, Brown and Company, Boston. 1982.

Wells, H.G. *28 Science Fiction Stories of H.G. Wells*. Dover Publications, Inc., New York. 1952.

Wells, H.G. *Mankind in the Making*. Chapman & Hall, London. 1911.

Wells, H.G. *The Short Stories of H.G. Wells*. Ernest Benn, London. 1927.

Wells, H.G. *The Undying Fire*. The Macmillan Company, New York. 1919.

West, Mae. *The Wit and Wisdom of Mae West*. G.P. Putnam's Sons, New York. 1967.

West, Nathaniel. *Miss Lonelyhearts*. Farrar, Strauss and Giroux, New York. 1971.

Weyl, Hermann. *Philosophy of Mathematics and Natural Science*. Princeton University Press, Princeton. 1949.

Weyl, Hermann. *Space, Time, Matter*. Translated by Henry L. Brose. Dover Publications, Inc., New York.

Weyl, Hermann. *Symmetry*. Princeton University Press, Princeton. 1952.

Weyl, Hermann. *The Classical Groups: Their Invariants and Representations*. Princeton University Press, Princeton. 1939.

Weyl, Hermann. *The Open World*. Yale University Press, New Haven. 1932.

Weyl, Hermann. 'A Half-century of Mathematics' in *The American Mathematical Monthly*. Volume 58. October 1951.

Whetham, W.C.D. *The Recent Development of Physical Science*. P. Blakiston's Son & Co., Philadelphia. 1904.

Whewell, W. *History of the Inductive Sciences*. Volume I. D. Appleton, New York. 1890.

Whewell, William. *The Philosophy of the Inductive Sciences*. Volume I. John W. Parker, London. 1847.

White, Arthur. 'Limerick' in *Mathematical Magazine*. Volume 64, Number 2. April 1991.

White, E.B. *The Trumpet of the Swan*. Harper & Row, New York. 1970.

White, William Frank. *Scrap-book of Elementary Mathematics*. The Open Court Publishing Co., Chicago. 1910.

Whitehead, Alfred North. *A Treatise on Universal Algebra*. Hafner Publishing Company, New York. 1960.

Whitehead, Alfred North. *An Introduction to Mathematics*. Oxford University Press, London. 1948.

Whitehead, Alfred North. *Modes of Thought*. At the University Press, Cambridge. 1938.

Whitehead, Alfred North. *Principia Mathematica*. At the University Press, Cambridge. 1925.

Whitehead, Alfred North. *Process and Reality*. The Macmillan Company, New York. 1929.

Whitehead, Alfred North. *Science and the Modern World*. The Macmillan Co., New York. 1929.

Whitehead, Alfred North. *The Concept of Nature*. At the University Press, Cambridge. 1930.

Whitehead, Alfred North. *The Organization of Thought*. William and Norgate, London. 1917.

Wiener, Norbert. *Ex-Prodigy*. Simon and Schuster, New York. 1953.

Wiener, Norbert. *I am a Mathematician*. Doubleday, Garden City. 1956.

Wigner, Eugene. 'The Unreasonable Effectiveness of Mathematics in the

Natural Sciences' in *Communications in Pure and Applied Mathematics*. Volume 13. February 1960.

Wilczek, Frank and Devine, Betsy. *Longing for the Harmonies*. W.W. Norton & Company, New York. 1988.

Wilde, Oscar. *The Picture of Dorian Gray*. The World Publishing Co., Cleveland. 1946.

Wilde, Oscar. *The Importance of Being Earnest*. Appeal to Reason, Girard. 1921.

Wilkins, John. *The Discovery of a World in the Moone*. London. 1638.

Willerding, Margaret F. 'The Uselessness of Mathematics' in *School Science and Mathematics*. Volume LXVIII, Number 6. June 1968.

Williams, Horatio B. 'Mathematics and the Biological Sciences' in *Bulletin of the American Mathematical Society*. Volume 1, Number 38. May–June 1927.

Winsor, Frederick. *The Space Child's Mother Goose*. Simon and Schuster, New York. 1958.

Wittgenstein, Ludwig. *Culture and Value*. The University of Chicago Press, Chicago. 1980.

Wittgenstein, Ludwig. *Remarks on the Foundations of Mathematics*. Translated by G.E.M. Anscombe. B. Blackwell, Oxford. 1956.

Wittgenstein, Ludwig. *Tractatus Logico-Philosophicus*. Translated by D.F. Pears and B.F. McGuinness. Humanities Press Inc., New York. 1961.

Wolf, Alan. *Parallel Universes*. Simon and Schuster, New York. 1988.

Woodward, Robert. *Probability and Theory of Errors*. John Wiley & Sons, Inc., New York. 1906.

Wordsworth, William. *The Prelude*. D. Appleton & Company, New York. 1850.

Wright, Frank Lloyd. 'The Architect' in *The Works of the Mind*. Edited by Robert Heywood. The University of Chicago Press, Chicago. 1947.

Zamyatin, Yevgeny. *We*. Translated by Mirra Ginsburg. The Viking Press, New York. 1972.

Zee, A. *Fearful Symmetry*. Macmillan Publishing Company, New York. 1986.

Zemanian, Armen H. 'Appropriate Proof Techniques' in *The Physics Teacher*. Volume 32, Number 5. May 1994.

Zukav, Gary. *The Dancing Wu Li Masters*. William Morrow and Company, Inc., New York. 1979.

PERMISSIONS

Grateful acknowledgement is made to the following for their kind permission to reprint copyright material. Every effort has been made to trace copyright ownership but if, inadvertently, any mistake or omission has occurred, full apologies are herewith tendered.

Full references to authors and the titles of their works are given under the appropriate quotation.

A MATHEMATICIAN'S APOLOGY by G.H. Hardy. Copyright 1947. Reprinted by permission of the publisher Cambridge University Press, Cambridge, United Kingdom.

AN INTRODUCTION TO PROBABILITY THEORY AND ITS APPLICA-TIONS by William Feller. Volume 1. Copyright 1960. Reprinted by permission of the publisher John Wiley & Sons, Inc., New York.

BLACK HOLES by Jean-Pierre Luminet. Copyright 1987. Reprinted by permission of the publisher Cambridge University Press, Cambridge, United Kingdom.

BUILDING MODELS BY GAMES by Wilfred Hodges. Copyright 1985. Reprinted by permission of the publisher Cambridge University Press, Cambridge, United Kingdom.

BYRON'S LETTERS AND JOURNALS edited by Leslie A. Marchand. Copyright 1974 by John Murray. Reprinted by permission of the publisher Harvard University Press, Cambridge, Mass.

CONVERSATIONS WITH JORGE LUIS BORGES by Jorge Luis Borges. Copyright 1968 by Jorge Luis Borges. Reprinted by permission of Henry Holt and Company, Inc., New York.

DEVELOPMENTS IN MATHEMATICAL EDUCATION by A.G. Howson. Copyright 1973. Reprinted by permission of the publisher Cambridge University Press, Cambridge, United Kingdom.

EINSTEIN: A CENTENARY VOLUME by A.P. French. Copyright 1979 by the International Commission on Physics Education. Reprinted by

SUBJECT BY AUTHOR INDEX

-A-

abstraction

Whitehead, Alfred North

...to be an abstraction..., 1

Mathematics is thought moving in the sphere of complete abstraction..., 1

add

West, Mae

One figure can sometimes add up to a lot, 4

addition

Carroll, Lewis

She can't do Addition..., 2

Dickson, Paul

Some of it plus the rest of it is all of it, 2

Dostoevsky, Fyodor Mikhailovich

Twice-two-makes-four..., 2

Hardy, Thomas

...surely two and two make four now?, 2

Ice-T

I write rhymes with addition and algebra..., 3

Shaw, George Bernard

At school I got as far as addition..., 3

Unknown

1 + 1 = 3, 4

If equals be added to equals..., 4

One and one make two..., 4

West, Mae

I learned that two and two are four..., 4

algebra

Barrie, James Matthew

Algebra! It-it is not a very ladylike study..., 5

What is algebra exactly..., 5

Brahmagupta

...the man of knowledge will eclipse the fame of others...if he proposes algebraic problems..., 5

Butler, Samuel

The clock does strike by Algebra, 5

Cajori, Florian

The best review of arithmetic consists in the study of algebra, 5

Carroll, Lewis

Let U = the University..., 6

Clifford, William Kinston.

...algebra, which cannot be translated..., 6

Cochran and Cox

...polynomials are notoriously untrustworthy when extrapolated, 6

Comte, Auguste

Algebra is the Calculus of Functions..., 6

Smith, Sydney
  What would life be without
    arithmetic..., 18
Unknown
  Arithmetically speaking, rabbits
    multiply faster than adders
    add, 18
White, E.B.
  The fifth graders were having a
    lesson in arithmetic..., 18
arithmetician
Steinbeck, John
  He was an arithmetician rather
    than a mathematician..., 18
asymptote
Frere, C.
  Where light Asymptotes o'er her
    bosom play..., 20

-B-
binary
Unknown
  In the binary system we count
    on our fists..., 21
binomial
Kaminsky, Kenneth
  ...go out in the hall just to use
    the binomial expansion, 22
binomial theorem
Unknown
  ...a man who may be a master of
    the binomial theorem..., 22

-C-
calculate
Kaplan, Abraham
  Come, let us calculate, 26
Nietzsche, Friedrich
  No more fiction for us: we
    calculate..., 26
Poe, Edgar Allan
  ...to calculate is not in itself to
    analyze, 26

calculation
Adams, Douglas
  It is known that there is an
    infinite number of worlds...,
    58
Babbage, Charles
  ...the erroneous calculation... be
    corrected as follows..., 23
Belloc, Hilaire
  The student must be careful in
    calculations involving the
    decimal point..., 23
Bennett, Charles H.
  Multiplication is vexation..., 23
Berkeley, Edmund C.
  The moment you have worked
    out an answer..., 23
Billings, Josh
  Tew kno exackly whare the
    sighn iz..., 24
Birns, Harold
  ...the following mathematical
    formula for bribery..., 125
Büchner, Ludwig
  Death is the surest calculation
    that can be made, 24
Cajori, Florian
  The miraculous powers of
    modern calculation..., 268
Churchill, Winston Spencer
  The human story does not
    always unfold like a
    mathematical calculation...,
    24
Cyrano
  Twice nothing is still nothing,
    267
Dickens, Charles
  With affection becoming in one
    eye and calculation shining
    out of the other, 24
Dirac, Paul Adrien Maurice
  ...I understand an equation
    when I can predict the

formulae

Kasner, Edward
There is a famous formula..., 74
Kipling, Rudyard
No formulae the text-books
know..., 74
Peirce, Charles Sanders
...a single formula without
meaning..., 75
Unknown
fo^r[M−u/l]a, 75

formulas

Dudley, Underwood
Authors who discover
formulas..., 74
Formulas should be useful, 74

fractions

Burr, Lehigh
...a quarter of eight is two, 76
Fosdick, Harry Emerson
...when life ceases to be a
fraction..., 76

functions

Gauss, Carl Friedrich
...functions...are only our own
creation..., 77
Keyser, Cassius J.
...given a transformation, you
have a function and a
relation..., 77
...the ordinary notion of a
function..., 77
McCormack, Thomas J.
...the notion of a function, 77
Roe, E.D, Jr.
The continuous function is the
only workable...function, 78
Unknown
...any student who had never
seen a Bessel function..., 78

-G-

geometer

Halsted, George Bruce

...such is the motto of the
goemeter, 84
Huxley, Aldous
...a god who 'ever plays the
geometer', 79
Scaliger, Joseph
A dull and patient intellect such
should be your geometers,
80
Unknown
Is it where the Geometer draws
his base..., 80

geometry

Adler, Irving
Geometry today consists of
many subdivisions, 81
Aristotle
We cannot...prove geometrical
truths by arithmetic, 81
Bell, Eric T.
...the efforts of beginners in
Geometry, 81
...the only royal road to
elementary geometry is
ingenuity, 82
This is the hog-tie, and it is what
Euclid did to geometry, 81
Berkeley, George
The method of Fluxions is the
general key by help whereof
the modern mathematicians
unlock the secrets of
Geometry..., 82
Bôcher, Maxime
...there is an independent
science of geometry..., 82
Borel, Émile
The goal of geometry is..., 82
Brown, Frederic
I've always been poor at
geometry..., 82
Cedering, Siv
Geometry indeed is God, 82
Chasles, Michel

...geometry often...gives a simple and natural way to penetrate to the origins of truths..., 83

Comte, Auguste
Geometry is a true natural science..., 83

Coolidge, Julian L.
...geometry is nothing at all, if not a branch of art, 83

Davies, Paul
Geometry was the midwife of science, 83

de Morgan, Augustus
Geometry is the application of strict logic..., 83

Dieudonné, Jean
...the art of geometry is to reason well from false diagrams, 84

Emerson, Ralph Waldo
The astronomer discovers that geometry...is the measure of planetary motion, 84

Euripides
Mighty is geometry..., 84

Fabre, Jean Henri
Geometry...the science of harmony in space..., 84

Galilei, Galileo
...the power of sharp distinction which belongs to geometry, 84

Hermite, Charles
I cannot tell you the efforts to which I was condemned to understand something of the diagrams of Descriptive Geometry..., 85

Herodotus
From this practice, I think, geometry first came to be known, 85

Hobbes, Thomas
...geometry, which is the mother of all natural science..., 85

Geometry...is the science that it hath pleased God hitherto to bestow on mankind, 85

Huxley, Aldous
...a world where beauty and logic,...and analytic geometry, had become one, 85

Klein, Felix
Projective geometry has opened up for us with the greatest facility new territories in our science..., 86

Kline, Morris
No branch of mathematics competes with projective geometry..., 86

Leibniz, Gottfried Wilhelm
...the whole matter is to reduce to pure geometry, which is the one aim of physics and mechanics, 86

Marvell, Andrew
As Lines so Love oblique may well/Themselves in every angle meet..., 86

Newton, Sir Isaac
...it is the glory of geometry..., 87

Pirsig, Robert M.
One geometry cannot be more true than another..., 87

Plato
...geometry will draw the soul..., 87
...the knowledge at which geometry aims..., 87
...we are concerned with that part of geometry which relates to war..., 87

Plotinus
Geometry, the science of the Intellectual entities..., 87

Poincaré, Henri
...geometry is not true, it is advantageous, 87

-H-

...a mathematical subject is in
danger of degeneration, 143
mathematical subjects
de Morgan, Augustus
The greatest writers on
mathematical subjects..., 131
mathematical talent
Dieudonné, Jean
...the flowering of mathematical
talent..., 131
mathematical theorems
Jevons, W. Stanley
In abstract mathematical
theorems..., 135
mathematical theory
Cayley, Arthur
...a mathematical theory...can be
perceived but not explained,
128
Newman, M.H.A.
That mathematical theory is a
lasting object to believe in
few can doubt, 138
Unknown
A mathematical theory is not
to be considered complete
until..., 143
mathematical thinking
Picard, Émile
Activity in mathematical
thinking..., 138
mathematical thought
Weyl, Hermann
The stringent precision
attainable for mathematical
thought..., 144
mathematical training
Huxley, Thomas
Mathematical training is almost
purely deductive, 135
mathematical truth
Colton, Walter
...the investigations of
mathematical truth..., 128
Lemoine, Émile

A mathematical truth is neither
simple nor complicated...,
137
Mill, John Stuart
...the evidence of mathematical
truths..., 223
mathematical work
Papert, Seymour
Mathematical work does not
proceed along the narrow
path of truth..., 138
mathematical writing
O'Brien, Katharine
Said an upside down A to an
inside-out E..., 148
Pétard, H.
...the following glossary is
offered to neophytes in
mathematical research..., 145
mathematical writings
Unknown
Trust me, it's true, 146
mathematically
Leibniz, Gottfried Wilhelm
All things in the whole
wide world happen
mathematically, 136
mathematicians
Adams, Henry
Every one except
mathematicians thought
mathematics a bore, 149
Mathematicians assume the
right to choose..., 149
Mathematicians practice
absolute freedom, 149
Adler, Alfred
...mathematicians cannot discuss
their mathematics at all, 149
Each generation has its few
great mathematicians..., 149
Perhaps mathematicians,
lacking the imagination to
appreciate the scope and

...I have thought of
mathematicians as
Rosicrucians..., 166

Stewart, Dugald
...I have never met with a mere
mathematician..., 166

Swann, W.F.G.
...a pure mathematician is never
as happy as when he does
not know what he is talking
about, 166

Swift, Jonathan
...the mathematicians of
Laputal..., 166

Sylvester, James Joseph
The mathematician lives long
and lives young..., 167

Synge, John L.
Mathematicians are human
beings, 167
The modern mathematician
weaves an intricate
pattern..., 167

Thom, René
...a mathematician should have
the courage of their most
profound convictions..., 168

Thompson, Silvanus P.
Do you know what a
mathematician is?, 168

Thomson, William [Lord Kelvin]
...nonsense about the
interference of
mathematicians..., 168

Tomlinson, H.M.
...but never that of the
mathematician, 168

Truesdell, Clifford A.
Now a mathematician has a
matchless advantage over
general scientists..., 168

Ulam, Stanislaw
Mathematicians, at the outset of
their creative work...,
169

Unknown
...an accomplished
mathematician, 174
...what the mathematician
predicts today..., 173
A binary mathematician..., 170
A doctor, a lawyer and a
mathematician..., 172
A mathematician is a device...,
170
A mathematician named Hall...,
170
A mathematician named Klein...,
247
A mathematician, a biologist
and a physicist are sitting...,
172
A mathematician, a physicist,
and an engineer were all
given a red rubber ball...,
170
A physicist, an engineer and a
mathematician..., 171
An engineer, a physicist, and a
mathematician are shown a
pasture..., 172
An engineer, a physicist,
and mathematician are
all challenged with a
problem..., 171
And God thinks she is a
mathematician, 248
He wasn't sure if he should
bet a mathematician, an
engineer, or an applied
mathematician, 170
Mathematicians are unable to
make the connection, 174
Mathematicians may flatter
themselves..., 174
Old mathematicians never die...,
170
One of the mathematicians
on the committee replied
dryly..., 174

-P-

parabola
Allen, Woody
　...and her figure described a set
　　of parabolas..., 288
Frere, C.
　...the fair parabola behold..., 288
Shaw, George Bernard
　..you will get hyperbolas and
　　parabolas..., 288
paradox
Bohr, Niels
　How wonderful that we have
　　met with a paradox, 289
Bourbaki, Nicholas
　...the major paradoxes which
　　provide food for logical
　　thought..., 289
Bunch, Bryan H.
　If we do not run into a
　　paradox..., 289
Eliot, George
　Play not with paradoxes, 289
Kasner, Edward
　Perhaps the greatest paradox of
　　all..., 289
Rogers, Hartley, Jr.
　It is a paradox in mathematics...,
　　290
Russell, Bertrand
　Although this may seem a
　　paradox..., 290
Smith, E.E.
　...any possible so-called paradox
　　can be resolved, 290
Thomson, William [Lord Kelvin]
　In science there are no
　　paradoxes, 290
Wilde, Oscar
　The way of paradoxes..., 290
parallelogram
Moultrie, John
　...forgetful of the claims of
　　curves and squares, and
　　parallelograms..., 291

Tolstoy, Leo
　...the diagonal of a parallelogram
　　of forces, 291
perfect number
Dickson, L.E.
　Perfect numbers have
　　engaged the attention of
　　arithmeticians..., 292
pi
Beckman, Petr
　The digits [of pi] beyond the
　　first few decimal places are
　　of no practical or scientific
　　value, 293
Burr, E. Scott
　To calculate pi to twenty-two
　　places, 296
Carter, Harvey L.
　A new value of pi to assign...,
　　293
Duffin, R.J.
　...give special prominence to His
　　favorite number, pi, 293
Fadiman, Clifton
　In order to remember the value
　　of pi to thirty places..., 296
Graham, L.A.
　A ring around the moon is π
　　times *D*..., 293
Graham. L.A.
　Trying to evaluate pi..., 294
I Kings 7 23
　And he made a molten sea...,
　　294
Kac, Mark
　...to explain why pi...keeps
　　cropping up in probability
　　theory..., 294
Kamath, B.A.
　To calculate pi to twenty-two
　　places, 296
Kirkpatrick, Larry D.
　To remember pi to twenty
　　places, 297
Morgan, Robert

Whitehead, Alfred North
The art of reasoning consists of getting hold of the subject at the right end..., 316

reasons
Shakespeare, William
His reasons are as two grains of wheat hid in two bushels of chaff..., 316

rectangle
Frere, C.
The sly rectangle's too licentious love, 317

recurse
Unknown
To iterate is human, to recurse divine, 318

recursion
Papert, Seymour
Of all ideas I have introduced to children, recursion stands out as the one idea that is particularly able to evoke an excited response, 318

recursive
Kelly-Bootle, Stan
See RECURSIVE, 318

referees
Lipschitz, R.
...a race of vampires, called referees..., 319

relations
Keyser, Cassius J.
To be is to be related, 320

research
Bates, Marston
Research is the process of going up alleys to see if they are blind, 321
Dodgson, Charles L.
Now this field of Mathematical research..., 321
Green, Celia
The way to do research..., 321
Lasker, Albert D.
Research is something that tells you that a jackass has two ears, 322
Mizner, Wilson
...if you steal from many, it's research, 322
Szent-Györgyi, Albert
Research means going out into the unknown..., 322
von Braun, Wernher
Basic research is when I'm doing what I don't know I'm doing, 322

-S-

set
Byron, Lord George Gordon
...the rest are but a vulgar set, 324
Cleveland, Richard
We can't be assured of a full set..., 324
Reznick, Bruce
A set is a set..., 325

set theory
Poincaré, Henri
Later generations will regard Mengenlehre as a disease..., 324
Quine, W.V.O.
Set theory is less settled and more conjectural..., 325
We may more reasonably view set theory..., 324

space
Lewis, Gilbert Newton
When we analyze the highly refined concept of space..., 276

sphere
O'Brien, Katharine
...at the sight of a cone with a sphere on top..., 326

square
Unknown

AUTHOR BY SUBJECT INDEX

Asimov, Isaac (1920–1992)
American author/biochemist
  mathematics, 179
  numerical conventions, 269
Asquith, Herbert (1852–1928)
English statesman
  hypothesis, 93
Aurelius, Marcus [Antoninus]
    (121–180)
Roman emperor/philosopher
  infinity, 99
  observation, 281
Auster, Paul (1947–)
American author
  numbers, 269
  prime, 303

-B-

Babbage, Charles (1791–1871)
English mathematician
  calculation, 23
Bacon, Francis (1561–1626)
English statesman
  mathematics, 179
  numbers, 270
Bacon, Roger (1220–1292)
English philosopher
  mathematics, 180
Baez, Joan (1941–)
US singer
  hypothesis, 93
Bain, Alexander (1818–1903)
Scottish philosopher
  mathematics, 180
Baker, W.R.
  dimension, 51
Banach, Stefan
  analogies, 12
Barnett, P.A.
  mathematics, 181
  reasoning, 315
Barrie, James Matthew
    (1860–1937)
English novelist
  algebra, 5

Barrow, Isaac (1630–1677)
English mathematical scholar
  mathematician, 150
  mathematics, 181
Barry, Frederick (1876–1943)
  hypothesis, 93
  mathematical, 124
Bartlett, Albert A.
  mathematics, 181
Bartlett, Elizabeth
  infinite, 99
Bates, Marston (1960–)
American zoologist/science journalist
  research, 321
Bateson, Gregory
  mathematics, 181
Baumel, Judith
  Fibonacci, 70
Beard, George M.
  error, 65
Beckett, Samuel (1906–1989)
Irish author
  triangle, 355
Beckmann, Petr
  pi, 293
Beerbohm, Max (1872–1956)
British author
  infinity, 100
Begley, Sharon
  mathematics, 181
  number theory, 270
Bell, Eric T. (1883–1960)
Mathematician
  common sense, 37
  ellipse, 61
  geometry, 81
  infinite, 100
  irrational numbers, 112
  line, 116
  map, 122
  mathematical reasoning, 124
  mathematician, 150
  mathematics, 181
  numbers, 270
  proof, 308

arithmetic, 15
proof, 309
Hubbard, Elbert (1856–1915)
US author/editor
reason, 315
Hubbard, John
mathematics, 209
Hudson, Hilda Phoebe
mathematics, 210
Hughes, Richard (1900–1976)
British writer
mathematician, 159
Hugo, Victor (1802–1885)
French poet/novelist
zero, 361
Huntington, E.V. (1874–1952)
order, 285
Huntley, H.E.
mathematics, 210
Huxley, Aldous (1894–1963)
English novelist
figures, 73
geometer, 79
geometry, 85
mathematics, 211, 334
number, 274
truth, 357
Huxley, Thomas H. (1825–1895)
English naturalist
hypothesis, 95
mathematical training, 135
mathematician, 159
mathematics, 211
mistakes, 261
prayer, 300

-I-
I Kings 7:23
pi, 294
Ice-T
US rap artist
addition, 3
Inge, William Ralph (1860–1954)
British Dean of St. Paul's Cathedral
mathematician, 159

Issigoinis, Sir Alec (1906–1988)
British automobile designer
mathematics, 211

-J-
Jackins, Harvey
equations, 64
Jacobi, Karl G.J. (1804–1851)
German mathematician
mathematics, 211
number, 274
Jeans, Sir James Hopwood
(1877–1946)
English astronomer
mathematical pictures, 135
mathematician, 160
mathematical, 211
models, 262
Jeffers, Robinson (1887–1962)
Poet
mathematics, 212
truth, 358
Jerome, Jerome K.
English writer
logic, 119
Jevons, W. Stanley (1835–1882)
British logician
deduction, 43
mathematical theorems, 135
number, 274
Johnson, Samuel (1709–1784)
English poet
algebra, 7
calculation, 25
numbers, 274
Jones, Cyrano
calculation, 267
Jones, Raymond F. (1915–?)
logic, 119
Jonson, Ben (1572–1637)
English Jacobian dramatist
observe, 282
Jowett, Benjamin (1817–1893)
British classical scholar
logic, 120